"Walters presents a compelling case that the 'deep ecological, demographic, and industrial roots' of these diseases must be considered if we are to minimize the danger of future emerging diseases."

—*Publisher's Weekly*

"This book is well written, thoroughly researched, and forces us to reexamine our role in preserving the environment."

—*E Magazine*

"In sharp, readable accounts of six recent 'plagues,' Walters points at the 1,000-pound gorilla customarily ignored in modern epidemiological discussion: underlying ecological causes. . . . He never rants, he is always calm, and he is scarily cogent."

—*Booklist Magazine*

"Walters' [book] . . . offers one of the most persuasive arguments I have read for treating the land and animals that inhabit it better."

—*The Seattle Times*

"In *Six Modern Plagues*, Mark Jerome Walters connects the dots between these phenomena and new epidemics of infectious disease. . . . Walters, a journalist and a doctor of veterinary medicine, manages to pack a ponderous load of fact and expert opinion into six concise, engaging chapters."

—*OnEarth Magazine*

"Mark Jerome Walters weaves a fine thread of human disturbances through the quilt work of modern pandemics. After being drawn engagingly into the explosive symptoms of global environmental change, readers will come to understand that we have no choice but to make peace with nature."

—PAUL R. EPSTEIN, M.D., M.P.H., Center for Health and the Global Environment, Harvard Medical School

Six Modern Plagues

Mark Jerome Walters

and How We Are Causing Them

Island Press / SHEARWATER BOOKS

Washington • Covelo • London

A SHEARWATER BOOK

Published by Island Press

Copyright © 2003 Mark Jerome Walters

First Island Press cloth edition, September 2003
First Island Press paperback edition, August 2004

SHEARWATER BOOKS is a trademark of The Center for Resource
Economics.

Library of Congress Cataloging-in-Publication data.

Walters, Mark Jerome.
Six modern plagues and how we are causing them / Mark Jerome
Walters.
p. cm.
Includes bibliographical references and index.
ISBN 1–55963–714–5 (alk. paper)
1. Epidemiology—Popular works. 2. Environmental health—
Popular works. 3. Human ecology—Popular works. I. Title.
RA653.W34 2003
614.4—dc22

2003015137

British Cataloguing-in-Publication data available.

Printed on recycled, acid-free paper ✺

Text design by Joyce C. Weston

Manufactured in the United States of America

10 9 8 7 6 5 4 3 2 1

To my mother, Antoinette

In memory of my father, Linwood

To my brothers and sisters
Andrew, Gregory, Patrick, Maryjane, George,
Christopher, Anthony, John, Maryann,
and Marylin

And in memory of my sister Toi

Contents

Introduction

I first learned of the strange new disease in the city while reading the *New York Times*. It was 1999, and just across the East River from my Manhattan office, several elderly victims had been admitted to Flushing Hospital Medical Center in Queens. They had been having trouble walking, were confused, and in some cases were comatose. Several soon died. Nearly a month passed before the affliction was identified as brain inflammation caused by an exotic virus. Before long we learned it was West Nile encephalitis, a disease originally seen in Uganda that was now being found for the first time in the Western Hemisphere.

Cases of the illness soon emerged near where I lived, in northern New Jersey, an hour's train commute from Manhattan. The idea that a potentially fatal disease almost unheard of there a few months before had suddenly popped up near my home was terrifying. Was this how the Black Death, which wiped out as much as one-third of Europe's population in the 1300s, or the 1918–1919 Spanish influenza epidemic, which killed at least 20 million people in my parents' lifetime, began? As a veterinarian, I am familiar with diseases, including some frightening ones. But no amount of medical training had prepared me for new, life-threatening diseases heretofore unknown in my neighborhood.

At the time, I wanted to dismiss West Nile virus as anomalous. Problem was, it wasn't the first new disease to appear during my lifetime or even in my town—nor would it be the last. Some outbreaks seemed like faraway curiosities, whereas others had become personal, everyday concerns. Lyme disease, which hadn't even been described until the mid-1970s, was now endemic in Morris County, where I lived. And then there was HIV/AIDS, a disease whose deadly global spread was known to almost everyone, not least of all those of us in the New York City region. Even mad cow disease and other afflictions I knew of only through the scientific literature sometimes seemed only a supermarket or an ill airplane passenger away.

More recently, in November 2002, a frightening new type of respiratory disease appeared in Guangdong Province, China, where several hundred people fell ill. This deadly and highly contagious pneumonia—it came to be called severe acute respiratory syndrome (SARS)—was rapidly spread by international air travelers, and within a month almost twenty countries, including the United States and Canada, were reporting cases. "We may be in the very early stages of what could be a much larger problem as we go forward in time," warned Julie L. Gerberding, director of the Centers for Disease Control and Prevention (CDC) in Atlanta, Georgia. Researchers in Hong Kong and at the CDC soon identified the infective agent as related to a family of viruses that cause the common cold. On the basis of genetic analysis, researchers concluded that the coronavirus probably came from a nonhuman animal. Although this was hardly a surprising finding to someone trained as a veterinarian, given that nearly 75 percent of new human diseases discovered over the past thirty years are carried by wild or domestic animals, I was no less worried. We acquired many ancient diseases from other animals, including smallpox from cattle and, apparently, the common cold from horses.

An enormous reservoir of potentially disease-causing viruses resides in wild animals, with many of these microbes remaining undetected until they suddenly appear on the human horizon. What's more, when a particular virus resides in humans *and* other animals, unlike one that exists only in humans, there is almost no way to eradicate it. The best we may be able to do is identify the animal reservoir and try to protect ourselves by showing a healthy respect for the natural boundaries between ourselves and that species. All this was not even to mention the super-exotic diseases such as Ebola hemorrhagic fever, which had undergone periodic irruptions among people and some wildlife in Sudan and Zaire during the previous two decades—deadly outbreaks that continue to this day. Infection with the usually fatal Ebola virus causes massive internal hemorrhaging. Barely a decade before West Nile virus broke out in New York, several monkeys infected with Ebola virus were imported into Virginia in what could have led to the first human outbreak of the disease in the United States. Fortunately, although several human handlers were infected, that particular strain of the virus proved innocuous in humans. Still, the point had been made: numerous new, sometimes fatal, infectious illnesses were pounding at the door. Some had already made it through.

But hadn't the surgeon general of the United States proclaimed, way back in the late 1960s, that the time had come when Americans could "close the book on infectious diseases"? Hadn't the miracle of modern medicine all but ended the war against pestilence?

In fact, now, more than three decades later, infectious disease still kills more than one in three people worldwide. The World Health Organization (WHO) reported in 1999 that "diseases that seemed to be subdued . . . are fighting back with renewed ferocity. Some . . . are striking in regions once thought safe from them. Other infections are now so resistant to drugs that they are virtually untreatable." Even the Central Intelligence Agency has expressed concern about the resurgence of infectious disease. In 2000 the CIA predicted that emerging infections will "complicate U.S. and global security over the next 20 years . . . endanger U.S. citizens at home and abroad, threaten U.S. armed forces deployed overseas, and exacerbate social and political instability in key countries and regions." This prediction was partially realized when, in April 2003, an estimated 10,000 residents of Chagugang, a two-hour drive from Beijing, rioted and gutted a building where SARS patients were supposedly to be housed. SARS riots elsewhere in China soon followed.

Scientists tell us that this global rise in infections comprises two general trends. Old diseases once believed to be controlled have resurged and in some cases have sprung up in new regions of the world. In recent years, malaria, an ancient disease, has dramatically increased in many areas, such as East Africa. This mosquito-borne illness kills nearly 2 million people annually. Half the victims are children under five years of age. Some forms of the disease have become resistant to chloroquine, a mainstay of malaria treatment. The disease is also appearing in places where it was supposedly eliminated. In 2002, a fifteen-year-old boy and a nineteen-year-old woman in Loudoun County, Virginia, contracted malaria from mosquitoes near their home—the first time in at least twenty years that malaria had been found in both humans and mosquitoes in an American community. In some areas of the globe, the increase in malaria has been linked to a warming global climate and degradation of forests, which have given mosquitoes more places to breed.

In 2002 the tropical paradise of Maui, Hawai'i, reported its first case of dengue fever in more than fifty years. Transmitted by a mosquito bite, this virus causes a sudden high fever, severe headaches, joint and muscle pain, vomiting, and rash. It is sometimes fatal.

Perhaps like many people, I was tempted to dismiss these

increases as artifacts of better detection methods. Weren't investigators simply picking up diseases that had eluded our older, cruder methods of surveillance? Unfortunately, the facts do not support this optimism.

A second, equally ominous trend is the emergence of new diseases, of which WHO has identified more than thirty just since 1980, including HIV/AIDS. Between 1980 and 1992, HIV/AIDS contributed to a 60 percent rise in fatal infections in the United States alone. Unknown three decades earlier, by 2003 AIDS had killed more than 20 million people and infected another 50 million worldwide. And the disease continues to spread. Tuberculosis, an ancient disease, kills 2 million people per year. Recently, however, new forms of TB have emerged that are resistant to at least two of the antibiotics traditionally used to treat the disease.

Lyme disease, first identified in 1976, is now the most common disease in the United States transmitted by a tick or other "vector." Then there are mad cow disease and its human manifestation, variant Creutzfeldt-Jakob disease (vCJD), which suddenly appeared in the United Kingdom in the 1980s. A host of lesser-known new diseases and infectious agents also contribute to WHO's list, including Nipah virus, toxic shock syndrome, the Kaposi's sarcoma virus, hepatitis C, and infection with a toxic strain of the *E. coli* bacterium. Although antibiotics, better sanitation, and other measures have lowered the percentage of deaths from infection worldwide since 1900, such improvements have hardly closed the book on infectious disease. If anything, we are in the process of writing entirely new volumes.

This emerging-disease phenomenon is actually more widespread than is at first apparent. Frogs and other amphibians around the globe have declined dramatically since the 1980s, partly because of novel infectious diseases. Plagues are striking a wide range of other species, including crayfish, seals, honeybees, wolves, gorillas, prairie dogs, ferrets, penguins, snails, snakes, wild dogs, salamanders, pelicans, and kangaroos, to name a few. Infections threaten to drive some species to extinction. Ebola hemorrhagic fever is rapidly wiping out many of the world's remaining wild gorillas. A cancer epidemic, apparently caused by a virus, threatens many species of sea turtle worldwide. Chronic wasting disease, a brain-destroying affliction similar to mad cow disease, is spreading among wild deer and elk in the western United States and could eventually spread throughout white-tailed deer in the East.

We've all heard some of these accounts, but our understanding tends to be based on piecemeal news, with little sense of an encompassing story. In some ways we are getting the least important part of the picture. Media reports usually describe isolated battles against new diseases and rarely tell us the larger ecological story of which many new afflictions are a part. The larger story is not simply that humans and other animals are falling victim to new diseases; it is that we are causing or exacerbating many of them, not least of all through the radical changes we have made to the natural environment. So closely are many new epidemics linked to ecological change that they might rightfully be called "ecodemics."

Intensive modern agriculture, clear-cutting of forests, global climate change, decimation of many predators that once kept disease-carrying smaller animals in check, and other environmental changes have all contributed to the increase. This is not even to mention increased global travel and commerce, which can rapidly spread many diseases. This view is not an alarmist's leap of the imagination; it is quickly gaining ground as evolutionary and epidemiological fact. Noted scientist Peter Daszak, executive director of the Consortium for Conservation Medicine in Palisades, New York, has put it this way: "Show me almost any new infectious disease, and I'll show you an environmental change brought about by humans that either caused or exacerbated it."

Environmental change and human behavior have long played a role in fostering epidemics. In fact, historians such as W. H. McNeill believe that major, extended waves of epidemics have swept across the human species on several occasions, beginning some 10,000 years ago, when the first agricultural settlements and close human contact with cattle and other livestock gave microbes a new bridge for jumping to humans, aiding the rise of smallpox, measles, leprosy, and other diseases.

Then, some 2,500 years ago, increasing contact among established centers of civilization opened new avenues for emergence or spread of disease, giving rise to a second extended wave of epidemics. Increased global exploration then ushered in a third phase of epidemics as indigenous peoples in Africa, the Americas, the Pacific region, and elsewhere fell victim to introduced diseases.

Mercifully, throughout the late nineteenth and most of the twentieth centuries, many societies enjoyed a dramatic decline in infectious disease. This was largely because a state of relative equilibrium had been reached: societies had developed immunity to many of these old diseases and had adjusted their ways of life to control them. Unfortunately, this period of relative microbiological peace has been short-lived; humans now appear to be entering a fourth phase of epidemics, spawned by an unprecedented scale of ecological and social change.

Although the scale of disruption may have changed, the underlying biological principles that often give rise to epidemics in times of instability have not: species compete for survival, and the contest is often between predators and prey. When disease-causing viruses and bacteria gain a decisive advantage over humans or other prey, sickness and even epidemics can occur.

Mammals—humans, wildlife, and domesticated animals alike—share much of the same "disease grid." Although the strains and types of disease-causing organisms may differ from one species to the next, they are often of related families. This means that the genetic changes needed for an organism to jump from one species to another may be small.

To survive in this competitive world, living things have evolved two basic reproductive approaches, which scientists have dubbed "r" strategies and "K" strategies. These approaches are geared to two different types of environments—a rapidly changing one and a relatively stable one. Faced with continual environmental unpredictability—droughts, freezes, hurricanes, fires, or floods, for example—an organism's best chance of producing future generations is in creating numerous young, jettisoning them into the world, and hoping for the best, so to speak. The more offspring they produce, the greater is the chance that at least some will survive. Survival may be a crapshoot, but an r-strategist species has almost unlimited rolls of the dice. Not surprisingly, the r strategy is exhibited by many small, rapidly reproducing organisms such as bacteria and viruses. The infectious agents that cause Lyme disease, HIV/AIDS, West Nile encephalitis, and numerous other illnesses, as we'll see, could all be considered r strategists. So are many of the insects or other vectors that transmit some of these diseases.

In contrast, larger animals tend to be K strategists. They produce few young and then try to nurture and protect them until they reach reproductive age. K strategists put their reproductive energies not so much into mass production as into caring for a few. Of course, such a strategy works well only in an environment that is reasonably predictable and stable. It would be wasteful, from a reproductive perspective, to put just a few offspring into a world so unpredictable that the parents' efforts to care for and protect them were likely to be overwhelmed by lurking dangers and sudden environmental shifts.

These reproductive strategies have enormous implications for the way we humans choose—or should choose—to treat the natural environments in which we live. As K strategists, we would be prudent to protect the relative ecological stability within which our species has evolved.

Of course, during this time, climate, land cover, and numerous other facets of our environment have naturally changed. But this has tended to happen slowly, over thousands or even hundreds of thousands of years, giving mammals the opportunity to slowly adapt. But many changes today are occurring over a single human life span. Like humans, many other large animals have evolved within and adapted to this long period of relative global environmental stability. And this is the state in which we K strategists find ourselves—or at least we did until recently.

Over the past century or more, humans have so disrupted the global environment and its natural cycles that we risk evicting ourselves from our shelter of relative ecological stability. This rapid change increasingly plays to the strengths of the r strategists, including our ancient evolutionary adversaries—the small profligate things such as bacteria and viruses whose reproductive strategy is well adapted to the very kind of environmental instability we are fostering. We humans are in the process of forfeiting our evolutionary home-court advantage. If the upsurge in new diseases is any indication, microscopic predators are taking full advantage of the instability.

Of course, in reality, ecological theories rarely play out neatly. Sometimes r strategists may opt for K strategies, and even what constitutes a radically changed environment may be up for debate. Nor are all r strategists bad. We and other mammals depend on the benign bacteria that inhabit our intestines and other parts of our bodies, displacing potentially dangerous ones. Bacteria help produce the soil we need for crops, and they serve numerous other essential ecological functions. Nevertheless, a few real-life examples suggest how ecological disruption can expand opportunities for disease-causing organisms at the expense of larger animals such as us.

Nipah virus is named after the place of its discovery in Malaysia in 1999. For humans, the infection is often mild, though it may elicit influenza-like symptoms, including high fever and muscle pain. In some instances, the disease progresses to encephalitis, or inflammation of the brain, leading to convulsions, coma, and, in half of symptomatic patients, death.

The natural reservoir of Nipah virus is the giant fruit bat of Southeast Asia, to which the virus apparently causes no harm. These enormous "flying foxes" normally feed in wild fruit trees in sparsely settled areas. But since the 1980s, logging and the spread of agriculture in Southeast Asia have decimated the bats' forested homes. In 1997 and 1998, human-set forest fires in Borneo and Sumatra, spurred by an El Niño–linked drought, blanketed much of Southeast Asia with a thick haze. This confluence of fires, unusual drought, and forest degradation caused the natural fruit crop to fail and forced the flying foxes to migrate farther north in search of food. They ended up in new locales, such as cultivated fruit orchards near pig farms in Peninsular Malaysia, where their hitchhiking virus attacked new species whose immune systems were unprepared for this microscopic invader. More than a hundred people died, and the Malaysian pig industry was devastated. Although reappearances of the virus in the region have been sporadic, the high fatality rate makes it a serious public health threat.

There are numerous other examples of sudden environmental change apparently benefiting a disease-causing microorganism. In the early 1990s a mysterious epidemic killed more than a billion fish along the North Carolina coast. Scientists eventually identified the agent of death as a dinoflagellate—a tiny aquatic organism—called *Pfiesteria piscicida*. The "cell from hell" was not a new creation, but the damage it caused almost certainly was. These researchers later discovered that when large schools of fish swam near toxic *Pfiesteria*, the dinoflagellates released a potent natural toxin. Marine biologist JoAnn Burkholder and her colleagues at North Carolina State University eventually showed that these *Pfiesteria* ambushes tended to occur in waters heavily polluted with waste from hog farms and other intensive agriculture in coastal watersheds. The microorganisms apparently capitalized on the polluted water by teeming in ever greater numbers into rivers and estuaries, where they attacked fish. Some fishermen, possibly exposed to toxin-laden aerosol on boats, reported dizziness and other problems after fishing near *Pfiesteria* outbreaks, and several of the scientists who cultured the *Pfiesteria* cells in fish in their laboratory aquariums fell seriously ill.

One of the more thoroughly documented associations between ecological change and the emergence of a human epidemic is seen in Lyme disease, named after the town of Old Lyme, Connecticut, where the disease was first described. Spread by the bite of an infected tick, the disease, if untreated, can lead to debilitating arthritis and other symptoms in people, horses, dogs, and some other animals. Lyme disease seems to be a classic case of humans destabilizing predator-prey relationships—with unintended results.

Rapid forest fragmentation and frequent logging in the forests of the eastern United States have led to an increase in mice and deer, two species that help transmit the disease. A dearth of foxes, owls, eagles, coyotes, and other predators—a consequence of habitat loss—has benefited these mice and other small rodents, and a lack of larger predators, including human hunters, has dramatically increased the number of deer. Housing developments within degraded woodlands have given the ticks more people to bite and infect. Warmer temperatures, on average, during the past three decades in parts of the eastern United States also may have helped the disease-carrying tick expand its range.

People in the United States, Europe, Japan, and other developed countries can no longer safely relegate new, exotic, or deadly diseases to faraway places. HIV/AIDS has invaded every part of the globe. The human form of mad cow disease infected people through meat served on ordinary dinner tables throughout England. In some regions, Lyme disease and West Nile encephalitis pose a risk to people in the city, in suburbia, or on a casual walk in the woods in spring or summer.

In the pages that follow, I tell of the human role in fostering modern epidemics through the stories of six modern diseases. Mad cow disease, first described in cattle in England in 1986, causes an ultimately fatal degeneration of the victim's brain. HIV/AIDS is moving toward the top of the list of humankind's most deadly modern scourges—with no end in sight. A form of food poisoning caused by a new strain of the salmonella bacterium known as DT104, which people usually contract by eating contaminated meat, is resistant to almost all the antibiotics commonly used to treat more traditional forms of salmonella food poisoning. The rate of Lyme disease is increased by ecological disruption of forests. A fatal illness caused by a newly discovered hantavirus, a virus spread by mice, makes regular appearances in the American Southwest and other areas, depending, in part, on global meteorological cycles. Unknown to almost all Americans just a few years ago,

West Nile encephalitis is becoming a dismal fact of life as the virus spreads throughout most of the United States. Six diseases, six parables of the unintended consequences of careless human disruption of the natural systems that are our home.

The ecological whole of these six diseases is far greater than the sum of their individual parts, and their significance is far greater than the relatively few people some of these diseases have affected. Together, these epidemics offer profound insights into the way we live, how we think, and the assumptions we embrace as children of the age of medical miracles. For all that modern medicine has prolonged life and relieved suffering, it has also fed the profoundly dangerous illusion that we are above or apart from the natural world, with its weather, forests, and cycles of life and death. The six diseases described in this book remind us that no amount of medical technology can rescue us from the fact— to state it simply—that we are K strategists increasingly slipping into an r strategist's world. We must once again face the heart-stilling discovery that human beings, as William H. McNeill put it, "will never escape the ecosystem and the limits of the ecosystem. Whether we like it or not, we are caught in the food chain, eating and being eaten. It is one of the conditions of life."

This realization is not all bad. In preserving the ecosystems on which health fundamentally rests, we stand to protect the health of many people for generations to come. In carelessly exploiting water, forests, fossil fuels, other species, and other natural resources, we will continue to sacrifice the long-term physical health of many for the financial gain of a few. Preservation of natural ecosystems, along with greater social equity, research, good surveillance, and benefits of modern medicine, can improve the health of not only people but many species. Human health does not belong to us alone. Nor, unfortunately, do the plagues we are all now experiencing.

The Dark Side of Progress:
Mad Cow Disease

1.

Near the village of Midhurst, West Sussex, an hour's journey south from London through green glens and soft hills, stands a seventeenth-century brick-and-timber farmhouse surrounded by purple hydrangeas and lipstick-red geraniums that tilt in the breeze. The lichen-covered clay tile roof and weathered walls seem to have grown from the earth itself. Sprays of red and yellow flowers spill from every corner of the grounds, and wild roses climb a trellis above a gate leading down to lush pasture and an ancient stone stable. It is as if Pitsham Farm were drawn from the enchanted poetry of William Wordsworth, where "majestic herds of cattle, free / To ruminate, couched on the grassy lea." Or so it might have seemed until, three days before Christmas in 1984, one of Peter Stent's cows began acting strangely.

"At first we dismissed her as a cow with a bad disposition, kicking in the milking parlor and all that," Stent told me. But when she got worse, Stent called his veterinarian, David Bee, who visited the farm. The cow hunched her back, leading Bee to believe she might have a painful kidney ailment. More cows soon fell ill, and Bee returned several times to attempt to diagnose the ailments. The first cow grew worse, developing head tremors and an unsteady gait. In February 1985 she died. The mysterious illness continued to spread through the herd. At a loss for a diagnosis, Bee dubbed the affliction "Pitsham Farm syndrome." Whatever the root of this malady, Bee concluded that it was attacking the brain, and he and Stent decided to ship a sick cow to the local agriculture ministry.

"I shall never forget that cow," Stent said. "The man came with a trailer already loaded with two sheep on their way to slaughter. When we prodded the cow into the trailer, she saw the sheep; then she went berserk and killed them. I thought she was going to destroy the whole trailer. She was extremely violent." Unfortunately, when the cow arrived at the local ministry she was killed with a gunshot to the head, which destroyed the brain and rendered it useless for analysis.

Determined to find the cause, Stent and Bee loaded up cow number 142—the tenth cow to be afflicted with the illness—and had her driven to the ministry. The head was removed intact and sent to the Central Veterinary Laboratory in Weybridge, Surrey, where the brain could be examined by a pathologist.

A stocky man with gentle blue-gray eyes, Stent sat in a lawn chair at a table and paused to sip his tea. A row of royal purple foxgloves nodded in an early summer breeze from the English Channel, twenty miles away. "Spooky behavior for these kindly animals," he recalled. "I cared about them and hated to see them sent to slaughter."

Stent's wife, Diana, appeared in the sunny yard and refilled our cups. A wood thrush sang three platinum notes followed by a reedy tremolo from a bush near an abandoned brick privy. Stent separated his right hand from his teacup long enough to make a short, sweeping gesture. "It's becoming more difficult to make a living from the farm anymore. I'm fortunate indeed to have other means. The price of milk has gone so low that we can't compete with larger operations. Now, with the Channel Tunnel open, tanker trucks bigger than my milking parlor bring cheap milk from the continent. We have 600 beef cattle, but people's feelings have really changed about eating meat."

Eager to give his respected veterinarian a place in our conversation, Stent called Bee on his cellular phone to arrange a meeting. We drove the backroads of the 600-acre farm past mostly empty pastures. When we arrived at the dilapidated milking parlor, Stent leaned out the car window and pointed inside the building's wide doorway. "I couldn't justify modernizing the operation in light of things. Now, look in there. Those are the old feed bins. At the time, I couldn't imagine what my cows were being fed. It's in there we first noticed the cows acting strangely. That's the spot where BSE began as far as the history books are concerned," he said, using the initials for bovine spongiform encephalopathy, the technical name for mad cow disease.

Bee's clinic was in the village of Liss, a twenty-minute drive. "As if BSE weren't enough, the foot-and-mouth epidemic last year finished off a lot of farms," Stent said as we drove along. He was referring to another highly contagious cattle disease that had recently swept through the United Kingdom. Although not dangerous to humans, it is one of the most contagious and economically devastating livestock diseases. "We didn't get foot-and-mouth at Pitsham, but we were quarantined like farms throughout the U.K. Any farm with the disease, all the animals were burned."

Bee greeted us in the waiting room of his clinic and ushered us into a treatment room so we could talk without interruption. A man in his late forties or early fifties, he wore wire-rimmed glasses, and his eyes shone with inquisitiveness. "Still haunts me sometimes," he said, recalling his first encounters. "You'd never recognize it in an undisturbed grazing herd. Then I'd walk up to the fence and suddenly a cow two hundred yards away would lift its head and fix its gaze on me with an eerie hypervigilance. 'That cow's infected,' I'd say to myself. If you stressed it, its symptoms could explode into kicking, tremors, aggression, a wobbly gait. An infected cow would come apart at the seams. Really spooky."

The task of examining cow 142's brain fell to Carol Richardson, a pathologist at the Surrey laboratory. She noted a strange sponge-like appearance strikingly similar to what is found in sheep with a well-known neurological disease called scrapie. Richardson wrote "spongiform encephalopathy" on the necropsy form and left the slide for Gerald Wells, her supervisor, to examine. Wells confirmed Richardson's diagnosis and filed the slide.

A year later, a cow from Kent developed similar symptoms; it became clear that the disease was not limited to Stent's farm. When this cow's brain reached Wells' laboratory, he discovered that it also had a spongiform encephalopathy. In 1987, fourteen months after Richardson's diagnosis of cow 142, Wells hailed his own discovery of "a novel progressive spongiform encephalopathy in cattle" and published a paper on his finding without so much as mentioning Richardson's diagnosis of the cow from Pitsham Farm—the first-ever documented case of mad cow disease.

Prior to Wells' 1987 publication, several cows at a farm in Malmesbury, Wiltshire, seventy-five miles east of Pitsham Farm, had also developed a fear of walking over concrete or venturing around corners. Some hung their heads low as if exhausted. Others developed a high-stepping gait in their back legs as if walking on hot pavement. Milk production dropped. Cows fell down and couldn't get up. The epidemic soon affected fourteen counties in southern England.

Although mad cow disease was apparently a new affliction, it belonged to a class of known brain-wasting diseases called TSEs, or transmissible spongiform encephalopathies. The name indicates that the diseases can be contagious and lend a spongy appearance to the brain, just as in the cow's brain Richardson had described. The first human TSE, Creutzfeldt-Jakob disease (CJD), was described in the 1920s. This degenerative disease leaves its victims, in the early stages, with loss of memory, unsteady gait, muscle spasms, and jerky, trembling hand movements. Another TSE, scrapie in sheep and goats, was scientifically described in 1936, although one of its symptoms—violent scratching to the point of mutilation—had been known for centuries. About a decade later, a TSE was identified in ranch-reared minks. In 1957, yet another human TSE, kuru, was identified in Papua New Guinea. Then, in 1967, chronic wasting disease was identified in some deer and elk in the western United States. BSE was officially added to the list of TSEs in 1987 with publication of Wells' paper. But mad cow disease was not just another TSE: never before had the affliction expressed itself in such a widespread outbreak. By 1988 more than 2,000 cows had been stricken, and in 1992 alone more than 35,000 cases of BSE in cattle would be reported. By January 1993 almost 1,000 new cases in cows were being reported *every week*.

"Incurable Disease Wiping Out Dairy Cows," proclaimed a headline in London's *Sunday Telegraph* in 1987: "A mystery brain disease is killing Britain's dairy cows, and vets have no cure." Farmers began to fear for their livelihoods and their rural traditions. But at least they were not fearing for their own lives—not yet.

2.

In October 1989 a report surfaced describing a woman, believed to be at least thirty-six years old, who had been diagnosed with Creutzfeldt-Jakob disease. That disease struck, according to one study, fewer than one in every 10 million people in Britain and Wales each year, and its usual victims were middle-aged or older people; the average age of victims at the onset of the disease was fifty-seven. CJD in the young—a teenager, for example—is so rare as to typically occur only once every twenty or thirty years. Conventional wisdom held that CJD was either inherited or contracted from contaminated surgical instruments, transplants, or cadaver-derived growth hormones once used to treat dwarfism. When it was learned that the young woman had been associated with a farm where mad cow disease was present, people began to wonder whether she had contracted her disease from an infected cow. This was dismissed by a government scientific committee, however, which concluded that "the risk of Bovine Spongiform Encephalopathy to humans is remote."

In August 1992 came the case of Peter Warhurst, a sixty-one-year-old dairy farmer at Meadowdew Farm in Simister, north of Manchester, who died of Creutzfeldt-Jakob disease. Warhurst had culled a "mad cow" from his herd three years before. The prestigious British medical journal *Lancet* described this as "the first report of CJD in an individual with direct occupational contact with a case of BSE." The report said that the case was probably a chance occurrence but raised "the possibility of a causal link." It was not a link the government wanted to hear about. Livestock is a mainstay of the United Kingdom's economy, and the stakes were huge.

Kevin Taylor, the government's assistant chief veterinary officer responsible for BSE control, publicly dismissed the notion of a link between mad cow disease and CJD, saying, "I don't think that a link between this case and BSE is even conjectural." This echoed repeated claims by British agriculture minister John Gummer that there was "no evidence anywhere in the world of BSE passing from animals to humans" and that "on the basis of all scientific evidence available, eating beef is safe." At a boat show in Ipswich in 1989, Gummer had vouched for the safety of beef, this time in a BBC television report that showed him helping his four-year-old daughter, Cordelia, chomp down on a beef burger nearly the size of her face. "When you've got the clear support of the scientists who deal with these matters [and] the clear action of the government, there is no need for people to be worried," he proclaimed, "and I can say completely honestly that I shall go on eating beef and my children will go on eating beef because there is no need to be worried."

But new cases kept emerging. In May 1993, Duncan Templeman, a sixty-four-year-old Somerset dairy farmer, came down with CJD. There had been three cases of BSE on his farm, and he was a beef-eater. Eight months later, in

January 1994, a third dairy farmer, from Just, in Cornwall—some of whose cows had also contracted BSE—entered a hospital with loss of memory and slurred speech. The fifty-four-year-old farmer soon became mute, and he died of pneumonia some months later. The *Lancet*, which reported the case, concluded that "the occurrence of CJD in another dairy farmer . . . is clearly a matter of concern." Although the report emphasized that the farmer might have contracted the disease from his cows, the government's Spongiform Encephalopathy Advisory Committee emphasized that he might *not* have and that the case therefore did not require "the Government to revise the measures already taken to safeguard public health against occupational and other possible routes of exposure to the BSE agent." But one member of the committee warned that, should a fourth case arise, the tide of probability would turn: farmers were probably catching CJD from their cows. In September 1995, a fourth ill farmer came to light. As if that weren't convincing enough, a rash of puzzling CJD cases had begun to occur in beef-eating young people not associated with farms.

In 1993 fifteen-year-old Victoria Rimmer of Connah's Quay, Deeside, came down with CJD—the youngest reported victim in Great Britain in almost twenty-five years. Victoria had been exceptionally healthy until May 1993, when she began losing weight, developed trouble with her vision, and soon became apathetic. A brain biopsy revealed spongiform encephalopathy. Her condition deteriorated. She had fits, her body twitched uncontrollably, and she went blind. According to her mother, the British newspaper *Today* reported, beef burger was Vickie's favorite food. Kenneth Calman, England's chief medical officer, countered, "No one knows what illness she is suffering from . . . there is no evidence whatever that BSE causes CJD." Victoria soon fell into a coma that lasted four and a half years, ending in her death.

The notion that CJD might be linked to a person's diet was not new, and the supporting evidence was as tantalizing as it was scant. In 1984 the *American Journal of Medicine* reported four cases in which individuals who commonly ate animal brains—those of wild goats, squirrels, and pigs—came down with CJD. The authors concluded: "Our case, along with experimental evidence for oral transmission of Creutzfeldt-Jakob disease and other spongiform agents, support[s] the hypothesis that ingestion of the infective agent may be one natural mode of acquisition of Creutzfeldt-Jakob disease."

After the identification of mad cow disease, not surprisingly, such speculation increased. In 1997 a neurologist at the University of Kentucky came across a CJD patient in Florida, a native of Kentucky who had a long history of eating squirrel brains back home—not an uncommon practice in rural parts of the state, where the brains are sometimes scrambled with eggs or put in a meat and vegetable stew called burgoo. The neurologist later discovered that all five patients of a neurology clinic in western Kentucky who were suspected of having CJD had a history of eating squirrel brains. The patients were not related, and they all lived in different towns, facts that minimized heredity or direct contact as a means of transmission. The study was widely reported in the media but criticized in the scientific community; for one thing, squirrels apparently don't get spongiform encephalopathy.

In 1998 the *Lancet* reported the intriguing case of a sixty-year-old man from Italy who was admitted to the hospital with muscle contractions, an unsteady gait, visual difficulties, and problems speaking. Two weeks after admission, he became mute and couldn't swallow, and several months later he died. The man, as far as anyone knew, had no unusual eating habits. But about the same time he was admitted to the hospital, his seven-year-old cat developed uncontrollable twitches and episodes of frenzy and hypersensitivity to touch. The cat grew progressively worse and soon was unable to walk. There was no evidence that the cat, which slept on the owner's bed, had ever bitten him. Analysis of cells from the man's and the cat's brains showed remarkably similar abnormalities. Either the man caught CJD from his cat, the cat caught it from the man, both were infected by a common source, coincidence led them to become infected independently, or the cases were simply misdiagnosed.

Epidemiologists rightly caution that for every victim of CJD who had eaten the brain of a wild animal, there were thousands of other people who had eaten the same thing without contracting the disease. Such is the slippery nature of anecdotal evidence. But it is also worth noting that of the thousands of people who may have eaten BSE-infected beef, only a select few contracted the human form of mad cow disease.

By the end of 1995, ten suspected cases had been documented in young people in the United Kingdom. Senior government officials continued to insist there was no link with beef. Even as official denials flew, several prominent scientists, including some government advisors, were preparing a paper for the *Lancet* that would confirm people's worst fears—that the cow disease and the human disease were connected—by acknowledging the "possibility that [these cases were] causally linked to BSE." Not until just before the study's publication in the April 6, 1996, issue did the British secretary of state for health, Stephen Dorrell, admit to the House of Lords that the ten young people probably were suffering from what had become known as variant CJD, the human form of mad cow disease. Researchers soon added physical evidence to the statistical case: the agent of mad cow disease in humans was indistinguishable from the agent that caused BSE.

Mad cow disease seemed like medical science fiction. One of humankind's most ubiquitous domesticated companions, the dairy cow, widely known for its gentle nature, and a frequent subject of poetry and painting, had delivered a ferocious new disease unto its keepers. No one could say how the cows had gotten it, but speculation soon shifted to their pastured brethren the sheep. It was one more connection in a strange set of circumstances that seemed to link sheep, cows, and humans in a bizarre and unprecedented web of affliction.

3.

Scrapie, the illness of sheep and goats that can cause the animals to madly scratch themselves raw, was first clinically recognized in Great Britain in 1732. An early description from Germany describes how suffering animals "lie down, bite at their feet and legs, rub their back against posts, fail to thrive, stop feeding, and finally become lame. . . . Scrapie is incurable. . . . A shepherd must isolate such an animal from healthy stock immediately, because it is infectious and can cause serious harm in the flock." The French term for the disease translates as the "malady of madness and convulsions."

Not until 1936 was scrapie proven to be infectious, though its origins remained a mystery. In 1966 researchers at Hammersmith Hospital in London suggested it was no ordinary infectious agent because, whatever it was, it possessed no genetic material, or DNA. It therefore was not a living agent at all. Researchers drew their dramatic conclusions from the fact that DNA is fragile and can usually be destroyed by ultraviolet light, heat, or chemical disinfectants. But the scrapie agent remained infectious even after prolonged boiling, exposure to the extreme dry heat of sterilization, blasting with high levels of ultraviolet radiation, or even soaking in formalin and alcohol. Scrapie thus joined the strange fraternity of infectious brain-wasting diseases caused by a nonliving infectious agent. These are the perfect agents of disease: you can't kill them because they're already dead. But scrapie would not be the last in its class.

In 1957 American scientist D. Carleton Gajdusek and an Australian colleague began investigating kuru, the fatal neurological disease that was killing the Foré people, an ancient tribe of about 15,000 in Papua New Guinea. The victims' brains looked so much like those of scrapie-infected sheep that in 1959 American veterinarian William Hadlow suggested the two diseases were the same. Like scrapie, kuru was infectious—in this case, it was passed through the tribe by the ritualistic eating of brains of the dead. A neuropathologist noted further that the brains of kuru victims looked a lot like those of CJD victims. If kuru looked like scrapie and CJD looked like kuru, then CJD looked like scrapie. These three brain-wasting diseases came to largely define TSEs, and for his work on kuru Gajdusek would receive the 1976 Nobel Prize in Physiology or Medicine.

But "TSE" was merely a descriptive term signifying something transmissible that made the brains of its victims spongy. It revealed little about the disease-causing agents themselves. In the early 1980s Stanley B. Prusiner, at the University of California, San Francisco, School of Medicine, sought to unravel the mystery. He proposed the heretical idea, for which he would later receive the Nobel Prize, that TSEs were caused by a special protein—that is, nothing more than one of the body's common molecular building blocks bound together in a lethal way. Unlike bacteria and viruses, these special proteins, which he called prions, do not reproduce—or at least not in the case of mad cow disease. Rather, once in the victim's body, they force normal proteins into abnormal configurations. Prions don't replicate; they enslave. The notion of an infectious protein was strange enough. That it could also be inherited, as some cases of CJD showed, revealed an entirely new and fearsome type of infectious disease.

Just as the brains of various TSE victims looked a lot alike, so did some of the symptoms. In fact, the human form of mad cow disease was almost indistinguishable from the better-known CJD. The most striking clinical difference between the two diseases was the age of the victims. The term "variant CJD," or vCJD, was soon coined to reflect the finer distinctions.

The knowledge that people could get the disease from eating parts of infected cows, though a landmark discovery, was only one element of the larger story. How, in the first place, had the cows gotten it?

Scrapie-infected sheep remained the top suspect. Yet for centuries cows in England had intermingled with infected sheep, and there was not a single documented case of a cow becoming sick from scrapie. Nor, in the more than 350 years that scrapie had been known in England, was a single case documented of a person becoming sick from the sheep disease. If mad cow disease did in fact come from sheep, why had it just recently begun showing up in cows, let alone in people?

Perhaps a random mutation of the scrapie agent had suddenly made it infectious for cows—and people. Or maybe mad cow disease had nothing to do with sheep. Perhaps a protein in a cow's brain had randomly mutated into a lethal TSE protein. Then again, conceivably mad cow disease had been around for a long time at such low frequency that it had never been detected in bovines, let alone in humans—until something happened to cause an explosive epidemic.

No one could say exactly what changes had caused the emergence of mad cow disease, but scientists soon began to wonder whether the intensive management practices in the production and husbandry of cows and sheep in the United Kingdom were responsible. Over the previous few decades, for example, as livestock production had intensified, many relatively small farms had been absorbed into huge industrial enterprises where livestock was treated like oil, natural gas, or any other commodity. The animals' natural needs for space, proper diet, and other comforts had been overshadowed by demands for greater efficiency and profit—but at an unexpected cost.

The fact that BSE seemed to be transmitted by consumption of certain parts of infected animals would have, under natural circumstances, prevented its spread between sheep and cows for the simple reason that these placid herbivores don't eat each other.

Or do they?

4.

Cows, sheep, and other herbivores evolved over millions of years to eat plants. Just about everything about them is geared to living in a world of greenery. Their teeth are designed for grinding tough plants, not for grabbing prey or cracking bones. Their large, padded lips help them grasp and pluck short grasses from the ground. Their broad cloven hooves help steady their weight on grass and soft earth, where lush forage is likely to be found. What's more, the bovine digestive system is designed to extract hard-to-get-at nutrients from grasses and other vegetation. Through grinding action and fermentation, their three "stomachs" break down and absorb the nutrients contained in their tough, lignin-based diet. Bacteria living in their gut are equipped to break down plant fibers. Cows, like all species, tend to function best within dietary boundaries drawn by evolution.

Violating such evolutionary boundaries can seem unnatural, if not disgusting. The term "rendering" is a euphemism for refining and repackaging animals' blood and guts into palatable feed for livestock. For example, for decades renderers in France routinely added human excrement to the mix, creating a high-protein feed supplement that was sold to livestock producers throughout Europe—a practice not stopped until the year 2000. Ignoring natural dietary boundaries of species is more than bad manners; it can also be bad for our health.

In the mid-twentieth century, meat producers realized they could save money if they recycled and sold the normally discarded by-products of butchered livestock, including the intestines, bladder, udder, kidneys, spleen, stomach, heart, liver, lungs, and other organs, as well as the bones. Through the process of rendering, these leftovers could be turned back into feed for cattle, sheep, and other herbivores. Nature's plant-eaters could be transformed into human-made carnivores.

The problem was that high-protein diets can cause serious problems in digestive systems designed for grass and other low-protein food. But the livestock producers saw this as nature's problem, not their own. Although cattle fed high-protein diets did routinely suffer from digestive problems, the animals usually survived to market, and, whatever the consequences for the animals, the effect on the profit margin was positive.

In the process of rendering, the use of heat, mechanical pressure, and chemical solvents reduces entrails and other organs into two basic chemical components. One product is fat, known as tallow, which is used for anything from soap manufacture and human consumption to production of animal feed and chemicals. The other product is greaves, used in fertilizer or as high-protein feed for cows, sheep, and other animals. Greaves can be further processed to yield a solid residue and small amounts of a valuable, highly purified fat used in perfumes and cosmetics. The solid residue can be ground up to produce concentrated meat and bone meal, or MBM. This is added to animal feed to boost the protein content, which can help the animals gain weight faster.

But scientists were puzzled. If rendering had caused mad cow disease, why had it not occurred forty years earlier, when rendering became a standard practice?

Although most prions survived rendering, one theory suggests that lowering the amount of heat or solvents in order to offset rising costs during the energy crisis of the 1970s allowed even greater quantities of the scrapie agent to remain intact. Therefore, more prions ended up in MBM and in the diet of cows. Or, possibly, the modified rendering process physically altered the agent, making it more infectious for cattle. But such changes in rendering during the 1970s had occurred throughout Europe, so why did mad cow disease emerge only in the United Kingdom?

A unique British contribution to the emergence of mad cow disease may have been the dramatic increase in the number of sheep in the United Kingdom about the same time the disease emerged—from about 31 million sheep in 1980 to more than 44 million in 1990. This in turn meant that a greater number of scrapie-infected sheep carcasses were being sent to rendering plants—and ending up as MBM. By 1985 there were about two sheep for every cow in England, which meant that cattle in England were probably eating more scrapie-infected sheep, via MBM, than anywhere else in Europe. Perhaps the increased number of infected sheep consumed by cows tipped the balance to an infective dose of scrapie. Or perhaps the greater number of scrapie-infected sheep simply increased the probability of a random change occurring in the infective agent, thereby turning a sheep disease into a bovine and human one.

"The most widely accepted hypothesis is that BSE originated in scrapie-infected sheep, but it's still just a hypothesis," Marcus G. Doherr of the Department of Clinical Veterinary Medicine at Switzerland's University of Bern, told me. "I don't think that riddle will ever be completely solved." Wherever it started, BSE rapidly spread in the cow population via feed containing meat and bone meal of infected animals. Whether farmers knew it or not—and Peter Stent was one who didn't—virtually all of them in the United Kingdom were feeding animal protein to their animals to make them grow faster.

Whatever the actual origin of the mad cow prion, intensive agriculture dramatically multiplied the agent and quickly sent it throughout the industrial food web. The British government's inquiry into the epidemic concluded that "BSE developed into an epidemic as a consequence of an intensive farming practice—the recycling of animal protein in ruminant feed. This practice, unchallenged over decades, proved a recipe for disaster." A 1988 ban on the feeding of recycled animal protein in the United Kingdom slowly stemmed the epidemic a few years later, but not before harm had been done to many humans and herds of cattle.

In January 1993, by the time the epidemic reached its peak, an estimated 1 million cows had been infected. By November 2000, cases had been confirmed in more than 35,000 herds in the United Kingdom. Cases also appeared in Belgium, Denmark, Switzerland, Italy, Greece, Germany, France, the Netherlands, Portugal, Ireland, and Spain. Before long Japan and Israel had cases as well.

By early 2002, a total of 125 cases of the human form of mad cow disease had been reported worldwide: 117 in the United Kingdom, 6 in France, and 1 each in Ireland and Italy. Almost all the victims had lived in the United Kingdom between 1980 and 1996, the time of the BSE outbreak. In fact, there has never been a human case in which the patient did not have a history of exposure in a country where the disease was occurring in cattle.

Fearing the disease might reach the United States, in 1997 the Food and Drug Administration banned the practice of feeding animal by-products to cattle. The FDA stated it was consequently "highly unlikely that a person would contract vCJD today by eating food purchased in the United States." Nevertheless, in May 2003 a bovine form of the disease came perilously close to the United States when a six-year-old cow tested positive for BSE in a 150-head herd in Alberta, Canada—the first case of a cow born in North American being afflicted.

5.

State of Colorado
Roy Romer, Governor
Department of Natural Resources
Division of Wildlife

20 December 1997
Christopher Melani

Dear Mr. Melani,

Thank you for participating in the Division of Wildlife's chronic wasting disease survey. I am writing to let you know that preliminary laboratory results indicate that the buck you harvested . . . was probably affected by chronic wasting disease. When it's convenient, I would enjoy speaking with you to get some additional information on the location where this buck was harvested and discuss any observations you may have made in the field. In addition, I would appreciate your marking the harvest site on the enclosed map and returning it to us for inclusion in our survey database.

Although there is no evidence linking chronic wasting disease to human health problems, I also want to advise you that the Colorado Department of Public Heath and Environment has recommended against consuming remaining meat from this carcass since laboratory tests indicate the animal may have been infected. If you wish to receive a refund on your license fee, please complete the enclosed application and return it to the Division of Wildlife. . . .

Sincerely,
Michael W. Miller, DVM, Ph.D.
Wildlife Veterinarian

The problem was that Chris Melani and his family had already eaten the venison. They'd also had part of it made into sausage, which they'd sent as presents to their friends.

Chris had shot the animal in the winter of 1997. As requested by the state, he sawed off the head and took it to the fish and game office, where pathologists would examine the brain for the telltale sponge-like appearance of TSE. The state encouraged the hunting of deer in affected areas, hoping to reduce the number of infected deer. If a brain was found to have TSE, the state would notify the hunter within ten weeks, with instructions to dispose of the venison.

"I didn't get a notice, so I figured everything was okay with the deer," Chris was quoted as saying, having waited several weeks without receiving word from the state. His wife, who also ate some of the venison, said, "What's done is done. You just go on with your life." So far, they are still healthy. Given the extremely long incubation time for the disease, many more years may pass before they feel truly safe.

First described in 1967 in captive mule deer in northern Colorado, chronic wasting disease, or CWD, causes infected deer and elk to develop a blank stare and to walk in repetitive patterns while their bodies waste away, though not until 1977 was the infective agent determined to be a prion. By the mid-1980s CWD had spread to new parts of Colorado and into

Wyoming. And despite the optimistic predictions of wildlife managers, by 2001 the disease had spread to Nebraska, to the Canadian province of Saskatchewan, and to captive herds in Oklahoma and Montana.

Even though there is no proven link between CWD and human illness, Chris Melani and his wife learned that three young venison eaters had in fact come down with a degenerative brain disease with symptoms almost indistinguishable from those of mad cow disease. Twenty-seven-year-old Jay Dee Whitlock II of Oklahoma, a truck driver, began having difficulty finding his hometown. Soon he forgot how to drive his truck. He died of vCJD on April 7, 2000. A hunter, he had eaten deer meat regularly for most of his life.

In May 1998, twenty-eight-year-old Doug McEwan of Syracuse, Utah, also an avid deer hunter, had trouble calculating his travel expenses after a routine business trip and then began forgetting the names of his wife and other close relatives and his home telephone number. He soon had difficulty talking, writing, naming objects, and dressing. He died ten months after the onset of his illness. His brain showed the telltale sponginess of vCJD.

About a year before McEwan fell ill, a twenty-eight-year-old woman showed up at the emergency room several times with weakness and difficulty in walking. When her condition grew worse, she was admitted to the hospital, where she died. A brain autopsy showed spongiform encephalopathy. As a child, she had regularly eaten deer meat harvested by her father.

Given that there are normally only five reported vCJD cases per billion people each year in people thirty years of age or younger in the United States, three in such a short time was more than worrisome. According to the Centers for Disease Control and Prevention, of the 4,700 reported "classic" CJD deaths in the United States between 1979 and 1998, the victims' median age was sixty-eight.

"You cannot say with absolute certainty that CWD won't transmit to people, but there is no evidence that it will," veterinarian Tom Thorne, director of the Wyoming Game and Fish Department, told me. Thorne, who has hunted and eaten venison from the infected area for thirty years without ill effect, calls himself the "longest ongoing experiment. All evidence indicates it doesn't jump from deer to elk to other animals, let alone to humans. You're more likely to get run over by a Winnebago in downtown Omaha."

Thorne emphasized that "the only thing that BSE and chronic wasting disease have in common is that they're both TSEs, and, possibly, they both originated from scrapie in sheep." Those similarities may be enough to stop some people from eating venison, but it obviously hadn't stopped Thorne. "CWD just isn't a risk to humans," he said. At the same time, Thorne acknowledged that it was "all a big mystery. We don't know where CWD came from. Maybe it was a spontaneous change in a protein that started in deer. Maybe they picked up the scrapie agent and it evolved to be pathogenic in deer. We just don't know." Pierluigi Gambetti, director of the National Prion Disease Pathology Surveillance

Center at Case Western Reserve University, who autopsied the brains of McEwan, Whitlock, and the other victim, was quoted as saying that he would not eat venison: "Why should I? I can eat something else. But that's not because I really think there is great danger. I just think the whole issue of prion disease in the United States, both in animals and humans, has to be confronted seriously." Although the brains of the young victims lacked the tell-tale physical signs common to TSE, five other vCJD patients—all at least middle age and who frequently hunted deer or elk—raised further suspicions of a possible link.

Chris Melani, his wife, and their friends, meanwhile, probably became a lot more worried when, in May 2000, Byron W. Caughey at the National Institutes of Health's Rocky Mountain Laboratories in Hamilton, Montana, demonstrated that chronic wasting disease, at least in the laboratory, is as infectious to human tissue as BSE. Then, in 2002, a cow was experimentally infected with CWD, suggesting one more worrisome similarity between the two diseases. That same year, both the FDA and the National Institutes of Health announced major new studies to determine CWD's contagiousness to humans and other species.

In 2002 the Wisconsin Department of Natural Resources also announced that three deer shot near the village of Mount Horeb had CWD—the first time the disease had been found east of the Mississippi River. Of the more than 500 deer later shot near Mount Horeb, 15 were infected.

Alarmed state wildlife officials began shooting deer from helicopters. Given that hunting is a multi-billion-dollar industry in some states, widespread infection among deer could deliver a severe economic blow. The density of deer in Wisconsin is high, and there is evidence that the higher the density of animals, the more deer become infected. According to Thorne, under some circumstances there is a 100 percent "attack rate," meaning that all the animals will come down with the disease and die from it.

CWD's appearance east of the Mississippi may mark the beginning of its inexorable march to the East Coast, where deer density is astronomical. The spectacle of blank-eyed, disoriented, and haggard deer wandering aimlessly through backyards in suburban New York and New Jersey would bring home awareness of the omnipresent web of infectious disease as never before.

6.

Today, more than six years after the discovery of mad cow disease in humans, memories of BSE still pulse through daily life at Pitsham Farm. "In retrospect, I'm appalled at what I didn't know about my own cows," Stent confesses. "I didn't know they were being fed other cows and sheep that had been ground into a powder. We've forced these hoofed grazers into cannibalism. On some farms they're fed growth promotants, and that's probably causing other problems. In many places in the world, livestock is kept in deplorable conditions, all for human convenience and profit. We've put cows on an assembly line and we take them off at the other end and butcher them. Did we really think we could just rearrange the world in any way we pleased? Nobody could have wished for or foreseen this awful thing called BSE. But should we be all that surprised?"

A Chimp Called Amandine: HIV/AIDS

Going up that river was like traveling back to the earliest
beginnings of the world, when vegetation rioted on the
earth and the big trees were kings. An empty stream, a
great silence, an impenetrable forest . . . the earliest
beginnings of the world.

—Joseph Conrad, "Heart of Darkness"

1.

It poured before daybreak, and by eight o'clock the smell of
raw meat hung in the humidity at the open-air market of
Oloumi. As I entered, the morning Air Gabon flight to Paris
was passing over Libreville. Inside the market, flanks of
duikers, smoked porcupine bodies, and crocodile tails covered the ground. A hairy arm languished in the shadows; a
leg was half hidden among brown fur bellies. A monkey lay
on its back in eternal slumber, its black, leathery palm half
open. My eye fell on an olive-colored tortoise leg, a giant

47

monitor's mottled tail, and the huge, mute beak of a hornbill. Eyes were everywhere, some ratcheted open in blank terror, others squinting as if into sunlight. There was the exhausted, depressed look of a palm-nut vulture and the stunned rosette face of a decapitated mandrill.

A man without a left forearm slid the limp black body of a cat-size colobus monkey off his right shoulder and onto the ground. An aged woman with dried blood on her forearms tossed it onto a pile of intact simian bodies behind her; then she slid another in front of her. Its hair had been singed off. She picked up a knife, carved through soft meat and the stubborn cartilage of a joint, and then loosely wrapped the severed arm and shoulder muscles in plastic and handed it to the man, who tucked it under his good arm and limped away, the casualty of some terrible misfortune.

One by one or in pairs, women with glistening, sunlit faces moved gracefully through the market. By midmorning the tables had been emptied of meat, and by noon Oloumi was quiet but for a few lingering souls packing up their remaining wares. I caught a taxi back to the hotel and ate bright orange carrot and parsley soup for lunch.

I had first come to Gabon in 1994 with colleagues from Tufts University School of Veterinary Medicine, near Boston, to help develop a plan for conserving that country's forests, which French timber companies were rapidly stripping away. Thousands of the loggers living in remote timber camps survived on bushmeat like that featured at Oloumi the morning of my first visit to the country, and the demand was draining the remaining forests of their wildlife, including gorillas, chimpanzees, and nearly every other kind of remotely palatable animal living there. At the time, the threat to wildlife brought about by such hunting was just becoming widely known, thanks to the efforts of the World Wildlife Fund and other groups that pressed for a reduction in logging, arguing that this could in turn reduce the demand for bushmeat.

Few people suspected at the time that the hunting trade might also be contributing to the emergence of AIDS. Whether bushmeat hunting was the vehicle that carried the virus from its source in wild animals to humans is still unproven, but mounting evidence points in that direction. Although the mystery of this elusive, ever changing, and still spreading virus remains, its message has become clear: human health does not exist apart from the larger natural world we share with other species. AIDS is not only a medical issue but also an ecological one.

2.

The conventional medical story of AIDS began in the late 1970s when doctors at three Los Angeles hospitals noted a cluster of illnesses among homosexual men. All the men had strange life-threatening infections usually limited to people with highly compromised immune systems. In 1981 the Centers for Disease Control and Prevention (CDC) published a report suggesting that a new immunosuppressive disease—apparently caused by a virus—had emerged; the following year the name "acquired immunodeficiency syndrome"—AIDS—was coined. Because the virus attacks the immune system, it leaves the body susceptible to other infections. Many erroneous theories—from nuclear fallout to drugs commonly used by homosexuals—arose to explain the malady's emergence.

Over the next few years, records came to light of immune-ravaged patients in Europe a decade or so earlier with symptoms similar to those of AIDS patients. All had a history of close links with Africa, which hinted that the virus may have emerged there. One of the early cases, for example, was that of a Portuguese taxi driver in Paris who had repeated bouts of illness. Several years earlier he had worked as a truck driver in Zaire, where he could have come into contact with prostitutes on the highway between Angola and Mozambique. Belgian physicians had begun to describe patients with similar symptoms in West Central Africa, but they were clueless about the cause.

In the 1990s the theory of the African origin of AIDS was strengthened when researchers in the United States discovered a frozen blood sample that had been taken from a Bantu man in 1959 for malaria research. The man was from Leopoldville, Belgian Congo, near the Congo River—now known as Kinshasa, Democratic Republic of the Congo. When scientists thawed and analyzed the sample, they found human immunodeficiency virus (HIV), the AIDS virus. This Bantu man currently remains the earliest documented case of HIV-1 infection.

But to know *where* HIV-1 had come from geographically

was not to know *what* it had come from. Was the virus new? If so, what had created it? Had it existed in some unidentified animal reservoir and, given the opportunity, jumped to humans, where it evolved into a deadly new disease? Or had the virus long lurked in humans, but with its spread limited by the isolation of villages where it initially occurred? Had modern transportation and large human migrations suddenly unleashed this long-established infection on the rest of the world?

Understanding the history of the virus, it turned out, was crucial to understanding its future, and while most scientists initially focused their attention on developing treatments for AIDS, a few pursued the lonelier quest to determine its origins. One of those was Beatrice Hahn.

3.

Born and educated in Germany, after medical school Hahn trained in the laboratory of Robert Gallo, the American co-discoverer of HIV. In 1995 she joined the faculty at the University of Alabama, where she worked to pinpoint the elusive origin of HIV-1.

"With the public health calamity of AIDS, it's not surprising that little money and effort was put into discovering the' virus' origins early on," she told me. We spoke in the summer of 2002 in her eighth-floor office at the University of Alabama at Birmingham Medical Center in downtown Birmingham. "But that was before people began to realize that understanding the origin of the virus might hold answers to controlling the disease. After that, the interest in discovering the origin of the virus really surged."

What made the quest complicated, Hahn said, is that the AIDS epidemic didn't stem from a one-time event. In fact, two different HIV families have been identified. "Evidence suggests that HIV-1-like viruses, which are responsible for nearly 99 percent of AIDS cases worldwide, have been introduced into humans on at least three occasions," she said. HIV-2, the second human AIDS virus, has jumped from animals to people on at least seven different occasions. HIV-2 is less virulent and is largely limited to parts of West Africa. But both types cause AIDS, and once these viruses entered the human population, they rapidly spread.

Exactly how they jumped is not clear. "We don't know if similar jumps are still occurring, but there's no reason to think they aren't. If anything, the opportunities for jumping have been increasing with the accelerating trade in bushmeat," she said, implying that there may be more HIV types out there that could infect people, if they have not already done so.

Hahn's description of her work brought to mind the image of a contemporary astronomer who tracks asteroids for a living. The chance of an asteroid striking the earth, we are told, is infinitesimal. But given enough asteroids and a long enough time, an impact is inevitable—with potentially catastrophic results. Hahn, however, is not looking at sterile asteroids from outer space. She is tracking evolving microbes from inner space—the space that separates humans from other species. She is tracking the origins of HIV by reconstructing its recent evolutionary tree.

A tall, intense woman with a German accent, Hahn cautioned: "You have to speak of HIV in the plural—HIVs. There is no one 'HIV virus.' There are many."

She interrupted our conversation to take a phone call and then stepped out of the office to have her secretary send a fax. She offered to get coffee, and when she returned a few minutes later, she rested her white mug on the knee of her jeans.

The closest thing to the human immunodeficiency viruses, Hahn explained, are the simian equivalents, SIVs. But "simian immunodeficiency virus" is a huge misnomer. The virus doesn't harm the immune systems of these natural hosts; they don't even seem to get sick. "If we could figure out how simian immune systems deal with the viruses, we might have a clue to controlling AIDS," she said. "That very hope drives my work."

Although immunodeficiency viruses were known in monkeys, not until the late 1980s was an SIV discovered that was almost indistinguishable from HIV-2. It came from a sooty mangabey, a small forest monkey of West Africa. That discovery strongly suggested that HIV-1 might also linger somewhere among the hundreds of other simian species in the forests of Africa. But exactly where, no one could say. Its eventual discovery would result from a combination of good science and good luck.

4.

In the mid-1980s, the National Institutes of Health sought to develop a vaccine to prevent HIV infection. The research would require chimpanzees as subjects. Chimps used in the study had to be screened to make sure they hadn't previously been inadvertently exposed to HIV through experiments that might have included HIV-contaminated blood. So the NIH began testing more than a hundred chimps held at the Alamogordo Primate Facility at Holloman Air Force Base in New Mexico. More than 99 percent were "clean."

The test results from a chimp named Marilyn, though, were quite different: her blood was full of antibodies to HIV or to something like it. Although the actual virus may or may not have been present in her body then—the test was incapable of showing this—Marilyn's immune system clearly had reacted to something like HIV. Where she had been exposed wasn't clear. Perhaps it was in the African forest of her birth, or perhaps it was during her captivity. It didn't seem to matter at the time; the main concern was that the positive test result had disqualified her from the NIH vaccine experiments. When the scientists who conducted the blood test published the results, though, they also warned keepers and researchers that chimps might carry HIV.

On December 17, 1985, Marilyn died from complications of childbirth. During a routine necropsy, tissue samples were collected and sent to Larry Arthur at the National Institutes of Health in Maryland, who froze them. That was the end of the story of Marilyn, the chimp who might or might not have had HIV. Or so it seemed.

5.

As Marilyn's tissues lay frozen at the National Institutes of Health, a common tragedy was befalling yet another chimpanzee in the forests of West Central Africa. A mother chimp was shot, perhaps in Equatorial Guinea or Cameroon; the location is uncertain. The hunter, or perhaps a middleman, brought the chimp's baby to Libreville and sold her to a childless French dentist and his wife. They named her Amandine. She was given a decorated room of her own in the couple's Libreville home, and in the family's photograph album she can be seen, dressed in toddler's clothes and sunbonnet, accompanying them on a skiing vacation in the Swiss Alps.

Amandine was often ill, in part because her diet lacked the fruits that chimps require. In 1988, her "parents" found Amandine clinging to the swing in her cage, screaming, her body rigid as if paralyzed. They uncurled her arms and lay the chimp prostrate. A physician in Libreville prescribed antibiotics, aspirin, and other medication. Over the next week, the episodes recurred. At one point, Amandine's left hand was clenched firmly closed, and she tried to pry it open with the fingers of her right. Her left leg and arm jerked. She screamed again and eventually went into a series of seizures. In desperation, her caretakers flew her to

Franceville, Gabon, to the primate facility Centre Inter-national de Recherches Médicales.

Robert Cooper, an American veterinarian, was the center's director of primatology at the time. "When Amandine arrived, she was one sick chimp. She was on the small side for a three-and-a-half-year-old. Her owners kept her in dia-pers, and she lived the life of anything but a chimpanzee. When she tried to stand up in our clinic, she fell over on the floor and started screaming," Cooper told me recently. A meticulous man, he kept copious notes from the time.

Cooper and a French colleague, veterinarian Jean-Christophe Vié, drew blood, testing of which revealed severe anemia. They gave Amandine a transfusion, and she im-proved. Vié and Cooper also sent a blood sample for routine HIV testing as part of a program that had been set up to screen wild primates. Cooper and the rest of the staff were stunned when the test came back positive. Unlike the test administered to Marilyn several years earlier, this one identi-fied the actual presence of HIV-1—or, rather, the simian equivalent, which was virtually indistinguishable. What's more, given the unlikelihood that Amandine had been infected by a human, she almost surely had been infected naturally in the wild. This suggested that she—and perhaps her kind—were the natural carriers of HIV-1.

Less than a year later, a second baby chimp with a remark-ably similar history was brought to the center for treatment. Her mother had been shot a few days earlier. Two French couples on a weekend outing happened to be passing through the village of Macolamapoye in northeastern Gabon when they came upon this chimp, who had an infected bullet wound in her arm. The couples took the chimp from her captors and made their way to the primate center in Franceville. GAB2, as the chimp was designated, died a few days later, but not before a blood sample was taken. By this time, Cooper had departed—fired, he claims, to clear the way for the Europeans who managed the center to claim ownership of the prized discovery—and Martine Peeters and her husband-to-be, Eric Delaporte, had assumed primary responsibility for the HIV-chimp study. When they tested blood from GAB2, it came back positive for HIV. Like Amandine, this animal had been infected in the wild. But the researchers' elation at possibly having discovered the origins of HIV-1 was tempered by a sobering thought: the couples who brought in the chimp, who happened to be their good friends, had had their arms and legs badly mauled by the chimp during the trip from Macolamapoye to Franceville. Saliva from the HIV-positive chimp had almost certainly entered their bloodstreams. Would they therefore contract HIV? Delaporte, a physician, was plenty worried. As one might imagine, so were the couples. Delaporte put them on sedatives while they anxiously awaited the results of their own HIV tests.

"The standard practice for accidental blood exposure to HIV was to give a drug called AZT, which was the only treatment available, and I recommended that all go on AZT as a precaution," Delaporte told me when I visited him at his office in Montpellier, France. "I got some sent from a physician friend in Paris, and we put them on it. Fortunately, all four of their tests came back negative.

"The fact that our friends didn't get infected doesn't mean that chimps can't transmit the virus through a bite. . . . Maybe it happens only one in a hundred times. We're just thankful it didn't happen in this case."

6.

The fortuitous discoveries of two HIV-positive chimps in Gabon were eureka moments in the quest to discover the origins of the disease, though these two cases were not in themselves definitive. In 1990, however, a third captive chimpanzee in Europe was found to be HIV-positive. Simon Wain-Hobson at the Pasteur Institute in Paris then used detailed genetic analysis to confirm that the virus fragments found in chimps were indeed related to HIV-1. And since their infections had occurred naturally, evidence for chimpanzees as the natural reservoir for HIV-1 was now rapidly mounting.

Then, as luck would have it, in early 1998 Larry Arthur at the National Institutes of Health telephoned Beatrice Hahn. "I'm cleaning out my freezer, and I've come across some old tissue samples saved from Marilyn," he told her. "Would you be interested in looking at them?"

"I jumped at the chance," Hahn told me. After analyzing Marilyn's thawed tissues using the latest technology, Hahn and her colleagues found evidence that the chimp had indeed been infected with an HIV-like virus—confirming earlier suspicions. Had Marilyn been artificially infected during experiments while in captivity, or had she acquired HIV in the wild? Hahn set about investigating Marilyn's past, examining the details of every experiment she had ever undergone—and there were dozens—to see if she could have been infected in captivity.

As far as anyone could determine, the chimp, perhaps orphaned by hunters, had been captured as a two- or three-year-old in 1963 and transported to the United States, where she may have spent some time at the Kansas City Zoo. In July of that year, Marilyn was moved to Holloman Air Force Base, which was beginning to collect chimps to use in tests for space flight. After carefully examining Marilyn's history of captivity, Hahn concluded that she had probably picked up her infection in the wild.

Hahn was ecstatic. The finding of yet another infected chimp bolstered her hypothesis. The evidence from Marilyn, Amandine, and GAB2—all belonging to the same sub-species—made a compelling case that the scourge of HIV-1 had originated in *Pan troglodytes troglodytes*.

Hahn hopes that the findings may one day provide clues for the control of AIDS. And how is it the chimps apparently remain healthy in the face of infection? Perhaps the ability of the chimps' immune systems to neutralize the virus will shed light on a human prevention. But Hahn's discovery also led to some larger personal insights. "When I began my work on the origins of HIV-1, I was a medical researcher," she said, shifting to a more contemplative tone. "When I learned that the virus came from chimps and saw how these animals were being slaughtered, it turned me into an unexpected conservationist. They are being hunted to the point of extinction, and any clues they may be able to provide could die with them. It doesn't matter if your goal is to protect public health or to protect endangered chimpanzees because the goals are one and the same. We're just not as separate from the animal world as we would like to believe."

7.

In order to explore further the theory that human handling of the meat of infected animals gave rise to AIDS, in 1998 Martine Peeters, Eric Delaporte, and their African colleagues began collecting hundreds of meat samples from markets in Cameroon and testing them for viruses. They found that more than 20 percent of the samples were infected with some form of SIV—an extraordinarily high rate of infection and theoretically enough to expose many people to the viruses every day—and they even discovered new SIVs in the process. Which ones are capable of infecting people, and under exactly what conditions they are most apt to do so, remains a mystery.

"When you go to one of these markets, you see a lot of the women with cuts and blood from the animals all over their arms and hands," Peeters said. "In theory, this is a perfect setup for transmission of the virus. Bushmeat is an enormous viral reservoir. There is a good chance there are many more unknown viruses in bushmeat, which is handled by numerous people, from hunters to dealers to the people who take it home. We have no idea what is actually being passed to people through blood and cuts during butchering or from bites from the animals or even their urine."

Peeters is not the only one who fears this unknown. As Harold W. Jaffe, an AIDS researcher at the Centers for Disease Control and Prevention, told the *New York Times*, "That is everyone's nightmare, that there is another virus out there that either could be or has been transmitted to humans that we cannot detect with current methods. No one wants to miss detecting the next HIV epidemic."

The concern persists because people are creating numerous new opportunities for the viruses to jump to humans by coming into ever more frequent contact with the blood and other body fluids of the simians that carry them. "The hunting of wildlife, which had always been an important source of subsistence food in the Congo Basin and throughout sub-Saharan Africa, has increased in the last decades," Peeters told me. "Commercial logging operations, many of them by European-based companies, have led to the building of roads in remote forests. This is followed by massive human migration and social and economic networks to support the logging. Several thousand people live in many of the logging concessions, and one of their main sources of food is bushmeat."

8.

Through the dusk and long into the night I waited on the cement platform at the train station in Lopé, an interior village of Gabon near the Réserve de Faune de la Lopé, a wildlife reserve where chimpanzees and other animals are regularly—and illegally—hunted. It was almost eleven o'clock, and the train to Libreville was already two hours late. Insects the size of hummingbirds buzzed the dim station lights. A freight train passed the platform, the cars fully loaded with tree trunks larger than the fuselage of a Boeing 737. I lay down on the warm cement, with a small suitcase under my head, and dozed. When I awoke twenty minutes later, a woman stood nearby keeping watch over two brightly colored nylon shopping bags with plastic handles. A bevy of flies clustered on the fabric. The woman swatted halfheartedly. When I stood up, I glimpsed the brown fur and black snout of a small animal protruding from the top of one bag. Perhaps this industrious woman was on her way to the market in Libreville to sell her bush meat.

Many hours late, the passenger train to Libreville finally emerged from the darkness. Over the next six hours, the train frequently shuddered to a stop and then began again on its whining, creaking way. At one point a young woman, assisted by a man, loaded three large sacks, one of them bloodstained burlap, into the coach. By daybreak the train was tunneling through walls of massive green trees and past wide rivers that stretched to the horizon. A light rain melted into forest mist, and the gloom of daybreak gave way to conversation and laughter.

The Travels of Antibiotic Resistance: Salmonella DT104

1.

On a cold early May morning in 1997, Cynthia Hawley stepped from her clapboard farmhouse in the green-tufted hills of Vermont's Champlain Valley and walked across the drive to feed the calves. To Hawley's dismay, her favorite calf, two-month-old Evita, who'd seemed perfectly healthy the day before, was listless, with sunken eyes and a grotesquely distended belly.

Experienced at treating her animals, Hawley gave Evita a shot of ampicillin and, with a needle, released the gas from her belly. Over the next hour, however, Evita grew worse. Hawley's veterinarian, Milton Robison, arrived and gave Evita some fluids for severe dehydration, more ampicillin, and an anti-inflammatory drug, but to no avail. At nine o'clock that night, her sweating head in Hawley's lap, Evita died.

By the next morning several more calves had fallen ill, and Robison returned. He now suspected infection with *Salmonella*, a genus of bacteria that occasionally causes diarrhea in entire herds. An ampicillin injection often solves the problem even when the infection perforates the intestines, invades the bloodstream, and turns the diarrhea bloody. But not this time.

Soon, 22 of the 147 cows at the 600-acre Heyer Hills Farm were ill, including several adults. Within days, 13 had died. Alarmed by the epidemic, Robison sent tissue samples from several of the carcasses to Cornell University's College of Veterinary Medicine in Ithaca, New York, for analysis.

The report from Cornell confirmed Robison's suspicion—but this was no ordinary salmonella. For one thing, analysis showed that the strain was resistant not only to ampicillin but also to four other antibiotics. Concerned that this profile might indicate a new strain, Cornell sent the samples to the National Veterinary Services Laboratories in Ames, Iowa. Word came back that the strain was *Salmonella typhimurium* DT104, a deadly variant that for more than three years had been haunting dairy farms—and people—in the United Kingdom and elsewhere. Hawley's sick cows were a prelude to only the third human outbreak of DT104 in the United States—and the first in the Northeast. The first, in 1996, struck nineteen schoolchildren in the small town of Manley, Nebraska, who were infected by drinking chocolate milk contaminated by infected cows. Fortunately, none of the children died.

Then, less than six months before Hawley's cows got ill,
there was an outbreak in Yakima County, Washington, and in
northern California, where more than 150 people got sick
from eating unpasteurized Mexican-style soft cheese. But
the disease was still unknown in the eastern United States.
In all her years working with dairy cattle, Hawley had certainly never seen or experienced anything like it. And she
herself had yet to become sick.

2.

The scientific history of *Salmonella* began in 1885, when
pathologist Theobald Smith isolated the organism in tissue
from pigs. But the bacterium was named after Smith's supervisor, veterinarian Daniel E. Salmon of Cornell University,
who had brazenly usurped credit for the discovery.

The term "salmonella" refers to a genus of bacteria that
lives in the intestines of many species. In many animals it
doesn't cause illness, but in others—including humans—the
sickness can be fatal. This depends in large part on the kind
of salmonella involved. *Salmonella typhimurium* and its different forms often cause food poisoning.

Some types of *S. typhimurium* evolved naturally. Others,
including those most dangerous to humans, are the accidental creations of our own practices. Strains of bacteria become
especially threatening when antibiotics no longer destroy
them. That is, the *Salmonella typhimurium* bacteria have
acquired antibiotic resistance. Although some bacterial
strains may resist antibiotics naturally, most become resistant

because of human overuse of the drugs. Through the phenomenon known as natural selection, bacteria frequently challenged by these drugs grow accustomed, in a sense, to their presence. Just as every human is different, so are most individual bacteria. And just as some people seem more resistant than others to a particular illness, so are some bacteria better equipped than others to survive in certain environments. When the immediate environment is awash in antibiotics, the vast majority of bacteria may die, but some of the better-suited ones will survive. This is the first step toward antibiotic resistance.

The surviving bacteria form the basis for the next generation and, of course, pass their traits to their offspring. Hence, more bacteria in the next generation will survive exposure to the same antibiotic. Through the process of natural selection, with continual exposure to drugs, each generation grows more resistant.

One environment where bacteria are frequently flooded by antibiotics is in large livestock operations, where producers frequently treat their cows and other animals with drugs to prevent epidemics in the unsanitary and overcrowded conditions commonplace in the industry. In the short term, it's cheaper to keep animals drugged than to keep them clean. Animals fed a steady diet of antibiotics with their grain also grow a little faster, thereby making the producers extra money. In addition, farmers often feed antibiotics to newborn calves—again, for the sake of short-term efficiency. These producers want to put the mother back in the milking parlor shortly after birthing, so they immediately send the newborns off to join thousands of other calves at a grow-out facility. Deprived of the natural antibodies in mother's milk, the newborns are given antibiotics to prevent infection.

Unfortunately, individual bacteria in the cows' intestines that survive this onslaught of antibiotics—and some almost always do—are highly resistant to antibiotics. Any number of genetic quirks, such as the particular makeup of the cell wall, for example, may let one bacterium survive where another would die.

The final step in human salmonella infection occurs when these resistant bacteria infect a person through undercooked meat or other contamination. Residual bacteria, which may include salmonella, often linger on the meat we bring home from the supermarket. Before cooking, when we open the package or prepare the meat, juices can contaminate the kitchen countertop and salad spinner, making it into refrigerated food and eventually into our mouths. Once taken into the stomach, the bacteria may pass into the intestines and enter the cells lining the intestinal wall, causing inflammation and pain. A fever may develop. In most healthy individuals, immune cells in the intestines will track down and kill the invaders, and the illness may pass as transient diarrhea. In severe cases, as in older persons, infants, or others with weak immune systems, or if the bacteria are especially virulent, the invaders perforate the intestines, causing bloody diarrhea and then escaping through the intestinal wall into the bloodstream. Thus, bacteria from a cow's stomach or intestines might normally give a person mild, transient diarrhea. If a bacterium happen to be a virulent one causing severe food poisoning, such as S. *typhimurium,* a short course of antibiotics usually cures the infection. But if the offending bacterial strain has already adapted to antibiotics in the farmyard, it could survive the treatment. A normally curable case of food poisoning could rapidly become fatal.

When a strain of bacteria is subjected long enough to a variety of antibiotics, very powerful offspring with a wide range of resistance are likely to develop. To make matters worse, a bacterium that grew resistant while living in a fish-farming operation may be picked up and carried by a bird, for example, to a stockyard, where that bacterium might trade parts with another and pass on its resistance in the process. Because the highly adaptive *Salmonella typhimurium* can infect birds, cattle, amphibians, and many other species, antibiotic-resistant illness can quickly spread across species. Not surprisingly, some of the most dangerous forms of salmonella have evolved in large-scale livestock operations. These epidemics often have begun with large producers and later infected smaller farms, such as Heyer Hills.

In today's world, bacteria travel the microbiological equivalent of the interstate highways—or, rather, international air routes or shipping lanes. In a world of incessant global commerce, there is no one to whom we are not ultimately connected. Where bacteria are concerned, Heyer Hills Farm may just be a shipment of cows or feed away from the United Kingdom.

3.

On Friday, May 16, Cynthia Hawley drove to Burlington, about twenty miles away. While waiting to see her hairdresser, Hawley felt a sharp abdominal pain, severe enough to make her double over. She lay down in a back room while her hairdresser phoned Hawley's sister and mother, who drove to Burlington to pick her up. Once she got home, the diarrhea and vomiting began. Over the course of that night, she grew weaker. The next day, with Hawley unable to retain any fluids at all, her mother insisted on driving her to Northwestern Medical Center in nearby St. Albans. By the time they arrived, she was unable to walk unassisted.

4.

DT104 was not the first drug-resistant form of *Salmonella typhimurium* to emerge from livestock agriculture. In retrospect, these earlier forms were a warning that business as usual would continue to create salmonella most unusual. Business didn't change; salmonella did.

The first major epidemic of drug-resistant *Salmonella typhimurium* struck the United Kingdom in the early 1960s. So-called type 29 defied antibiotics on a scale unprecedented at the time. As tracked by London's Public Health Laboratory Service, by 1963 type 29 had become resistant to two antibiotics commonly used in livestock. It soon picked up resistance to tetracycline, and by early 1964 it had added two more antibiotics to its list—to be followed a few months

later by two more. By 1965 more than 95 percent of salmonella type 29 tested in cattle in the United Kingdom showed
antibiotic resistance, and some rare forms had armed themselves against seven antibiotics. Of some 500 confirmed
human cases of salmonella type 29 food poisoning in the
United Kingdom, 6 were fatal. The epidemic's origin was
traced to a livestock dealer who, despite heavy use of antibiotics, had sick calves—and was selling them throughout the
United Kingdom. After being charged by the government
with illegal sale of sick calves, the dealer, a Mr. Atkinson,
apparently committed suicide by slamming his car into a
tree.

In the mid-1960s, E. S. Anderson of the Public Health
Laboratory Service concluded that the outbreaks were
"almost entirely of bovine origin" and warned that "the time
has clearly come for a re-examination of the whole question
of the use of antibiotics and other drugs in the rearing of livestock." An editorial in the British magazine *New Scientist*
argued that use of antibiotics to make animals grow faster
"should be abolished altogether."

With physicians fearing the same could happen in the
United States, a 1968 editorial in the *New England Journal
of Medicine* warned that antibiotics could become useless,
sweeping away a major modern line of defense against
infectious illness. Even common and treatable illnesses
such as pneumonia could produce vast epidemics with
numerous deaths. "Unless drastic measures are taken," the
article said, "physicians may find themselves back in the

pre-antibiotic Middle Ages in the treatment of infectious diseases."

The Food and Drug Administration agreed that there was cause for alarm and said that "there is ample data now in the literature to support more rigid control of antibiotics in animal feed and water." As C. D. Van Houweling, a veterinarian and chairman of the FDA's task force on antibiotics, explained, indiscriminate antibiotic use "favors the selection and development of single- and multiple-antibiotic-resistant bacteria . . . and could produce human infection." The logic applied to many kinds of bacteria. The FDA concluded in the early 1970s that at least licenses for use of antibiotics as growth promotants should be revoked.

The drug industry immediately launched an assault, in the media and even in scientific journals, on the FDA's conclusions. In a 1973 article published in *Advances in Applied Microbiology*, Thomas H. Jukes, a former biochemist for Lederle, one of the first commercial producers of antibiotics for livestock, blamed the FDA's conclusion partly on "a cult of food quackery whose high priests have moved into the intellectual vacuum caused by rejection of established values." He cited as evidence of the cult two bills then before the United States Congress that would "authorize definitions for 'organically grown food which has not been treated with preservatives, hormones, antibiotics or synthetic additives of any kind.'" Jukes also advocated that antibiotics be routinely used in some human food. "I hoped that what chlorotetracycline did for farm animals it might do for children," he wrote, referring to less disease, fewer illnesses and deaths, and "slight to moderate increases in growth." Antibiotics, he believed, could compensate for malnourishment and for the overcrowded and often unhygienic living conditions of many people in the developing world, just as they did for cattle. "This sounds like the conditions under which chickens and pigs are reared intensively," he wrote, concluding that similar benefits could result for humans.

Although the United States took no action to limit antibiotic use in livestock, in 1970 the British Parliament banned the use of almost all antibiotics to promote growth, despite fierce opposition from livestock producers and the drug industry. Over the next six years, the incidence of salmonella type 29 declined, presumably because the reduction in antibiotic use permitted populations of nonresistant forms of the bacteria to build up again. In other words, in the absence of antibiotics, resistant bacteria had no survival advantage over nonresistant ones. Unfortunately, the legislation did not restrict new antibiotics that would soon come on the market. A decade later, use of these new antibiotics in livestock would precipitate another salmonella epidemic—a second warning that continued routine use of large amounts of antibiotics posed a grave public health threat.

From about 1973 to about 1980, new types of resistant salmonella began to appear among cattle in the United Kingdom, largely in response to the routine use of antibiotics in animals—including a new multiple-drug-resistant strain that struck more than fifty farms in southern England and spread to Cambridgeshire and Yorkshire. Years later, the folly of the situation was summed up in testimony to a House of Lords committee investigating antibiotic use. The way antibiotics were being used, the witness said, reminded him of "the man who threw himself out of the Empire State Building and as he passed each window he said, 'So far so good, so far so good!'"

The National Office of Animal Health (NOAH), the representative of animal drug manufacturers in the United Kingdom, argued that there was no problem with continued routine use of antibiotics in livestock because "new antibiotics are being developed all the time." As a resistant bacterial strain developed, so the argument went, the industry would develop a new drug to counter it. NOAH did not point out that the "new" antibiotics were mostly spin-offs of existing ones, and therefore the bacteria would very likely be as resistant to them as to their immediate antecedents. The discovery and development of new antimicrobial drugs for multi-resistant organisms would, in fact, soon begin to slow as companies shifted their dollars away from research on new antibiotics and toward pharmaceuticals that carried a higher profit margin, such as cancer drugs.

In response to continuing salmonella epidemics in the United Kingdom and the United States, in 1977 the FDA actually proposed revoking the license for use of certain antibiotics as growth promotants for livestock. Congress buried the FDA's recommendation, however, by requiring the agency to take no action until the National Academy of

Sciences' National Research Council (NRC) completed a report on agricultural antibiotics. By that time, no fewer than half a dozen weighty scientific evaluations had already been completed, with the broad consensus in the primary scientific literature that overuse of antibiotics in agriculture posed a health risk to humans. The NRC report, however, concluded that since there was not *absolute* proof—no smoking gun—the practice should be continued for its economic benefits. The NRC study also pointed out that it was virtually impossible to definitively link antibiotic use in animals to food poisoning in humans caused by drug-resistant bacteria. This is because evidence of transmission is the contaminated food itself, which has usually been disposed of before people get sick and an investigation is begun. The report's encouragement of routine use of antibiotics was hardly surprising, given that the committee that wrote the report was chaired by Raoul Stallones of the University of Texas School of Public Health, a paid consultant to several animal-drug companies and an outspoken advocate for unlimited antibiotic use in livestock. Some time after release of the NRC's report, Stallones wrote, "If the decision were mine, the hog farmers could use all the antibiotic drugs they wish to make the pigs grow."

In 1982 a smoking gun was unexpectedly found when researchers at Harvard Medical School traced tetracycline-resistant illness in humans to tetracycline use in animals. By using genetic fingerprinting to exactly match the bacteria in the livestock with the bacteria in the patients, Thomas

O'Brien and his colleagues had, in effect, solved the case without using food as the witness. The FDA considered the study definitive, concluding that the issue "certainly has been studied sufficiently" and that no further evidence was needed to justify limiting or banning the use of certain antibiotics in livestock. Van Houweling, who had argued in favor of an FDA ban a decade earlier, meanwhile had become a consultant to the hog industry. In response to the FDA's latest conclusions, he made an about-face, stating that "history has shown that it doesn't make that much difference" if the drugs are banned in feed. Britain's ban of certain antibiotics in 1970, of course, had suggested exactly the opposite.

In the face of continued congressional opposition to limiting antibiotic use in livestock, the FDA took no action on conclusion of harm. The agency's budget, after all, was in the hands of the same appropriations subcommittee that handled the budget of the U.S. Department of Agriculture, which was heavily influenced by agricultural interests.

5.

"What brings you here?" the attending physician asked when Cynthia Hawley finally reached the emergency room.

"Acute gastroenteritis," she groaned, accustomed to using medical terminology on the farm. Remembering the veterinarian's warning that the infection was highly contagious to humans, she added: "The cows have it. It may be DT104."

That physician had never heard of DT104. The following

day, the case was assumed by Mara Vijups, a physician trained at the University of Vermont, who'd never heard of it either.

"Cynthia was medically off the charts," Vijups told me when I visited her clinic in Vermont. "I'd never seen anyone that sick from salmonella. Her blood count showed her blood was very toxic. Her face was gray. She was losing massive amounts of fluid through bloody diarrhea. It was all we could do to keep her alive that first night. She hadn't eaten in days; I really feared we'd lose her. We pulled up some articles on DT104, and I called Cynthia's veterinarian and the state veterinarian, since they had experience with it in animals. They warned me of what to look out for. I was very scared."

Vijups was used to having several drugs at her disposal to treat patients with severe salmonella infections. Ampicillin almost always worked, but as Vijups knew from her crash course in DT104, that drug would be powerless in this case. A combination antibiotic known as Bactrim was another option, but Hawley, like many people, was allergic to it. That left Vijups to ponder two remaining life-or-death options for her patient. One was cephalosporin. Although the drug was often prescribed for salmonella food poisoning, its effectiveness against the infection had not been widely studied, leaving open the possibility of unexpected failure in the face of DT104. The second option was fluoroquinolone, which had a long and distinguished track record against more traditional forms of *Salmonella typhimurium*. But DT104 had begun showing signs of fluoroquinolone resistance in the United States and Europe. Still, it was the best hope, and Vijups decided to prescribe it—and pray for the best.

6.

Several years before Hawley was infected with DT104, major outbreaks in Great Britain marked the third great salmonella epidemic that country had faced in less than thirty years. It wasn't that salmonella had a particular affinity for the English or Scottish countryside; the United Kingdom's surveillance system was among the best in the world. Many countries had salmonella epidemics without ever realizing the cause.

From the very first documented human outbreak of DT104 in Great Britain—when it struck seven people in Airdrie, Scotland, including five from one family—the signs were frightening. Not only did DT104 seem to kill more cows and make people much sicker than was the case in early *Salmonella typhimurium* epidemics, but even its physical design set it ominously apart.

The particular part or segment of a bacterium that makes it drug-resistant can occur in different places on the microbe. Usually a tiny segment on the bacterium neutralizes the antibiotic by either breaking it down or preventing the bacterium from ingesting it. These tiny bacterial segments can be shed or acquired as environmental conditions require. That is why suspending the use of certain antibiotics can cause resistant bacteria to slowly lose their resistance—and why prudent use of the drugs can possibly restore their curative powers.

But DT104 carried its resistance in a more or less permanent form—that is, within the cell's genetic material. Hope of its shedding the resistance, even in the absence of antibiotics, was lost. Once DT104 became resistant to a drug—and the DT104 that struck Hawley's farm had already become resistant to at least five of them—those drugs would probably remain impotent against the strain. The bacteria had mutated, and as a consequence resistance had become essentially permanent.

DT104 also differed profoundly from earlier drug-resistant salmonella in having apparently acquired its resistance from elsewhere. That is, even though the bacterium was first detected in the United Kingdom, its resistance seems to have come from far away. Exactly where the resistance genes originally came from no one could say. But Frederick J. Angulo of the Centers for Disease Control and Prevention was on their trail. He heads the CDC's National Antimicrobial Resistance Monitoring System (NARMS), whose task it is to identify and track dangerous characters of the microbial world—such as DT104—as they move through the United States.

Angulo is a cheery middle-aged man with a serious focus when it comes to bacteria. "In 1996 a CDC colleague in Geneva, Switzerland, e-mailed me an article about a bacterium that was isolated from a kitten in England," he told me when I visited his office in June 2002. "We didn't know at the time it had already reached the U.S. The appearance of DT104 was remarkable. It didn't slowly move from one country to the next, leaving a trail of intermediate forms as it evolved. The complete bacteria just exploded globally all at once, including in the United Kingdom and western Europe, Japan, and other countries.

"The first documented human infection in the U.S., it turns out, had been a man from Kansas in 1985. However, we did not become aware of the problem until the mid-1990s. The case in Kansas suggests that the bacteria had been lurking around, perhaps in cattle, in this country. It just took the bug a while to cross over to people through food. I originally thought DT104 had emerged in livestock agriculture, and that may still be the case, but there is some evidence that the resistance package of DT104 came not from livestock but from farms of a different sort—fish farms.

"Three of DT104's most distinguishing traits are directly related to its resistance," Angulo said. "Two of the particular resistances the salmonella carries are quite rare. In some cases, the only other place that occurred, and that was a few years before the appearance of DT104, was in bacteria that lived on farmed fish in Southeast Asia."

"How," I asked, "would the packet of resistance genes have moved from fish bacteria to *Salmonella typhimurium*?" Angulo didn't know for certain, but he suggested several plausible scenarios.

According to Angulo, an S. *typhimurium* bacterium

probably would have had to encounter a fish bacterium and pick up its resistance, perhaps in a pond filled with waste from both cattle and fish farms. Or a bird infected with the microbe could have visited an aquaculture facility and defecated in the water. "Wherever it happened, once salmonella had picked up the resistance, it could have gotten into fish meal made from discarded fish products. Fish meal is a common supplement in cattle feed. Contaminated feed could have rapidly been shipped to Europe, Japan, and the United States, where it then infected cattle in those countries. Then it was just a matter of time before it jumped to people."

The idea of bacteria being spread around the world in animal feed is not far-fetched. In the 1970s, a rare type of salmonella that caused outbreaks around the world was traced to fish meal from Peru. Fishermen had dried the fish on the decks of their ships, and seabirds infected with the salmonella defecated on the fish during the drying process. The bacteria became part and parcel of the fish meal, which was quickly spread through international trade. The fish meal was fed to poultry, the meat of which infected people. Although the bacteria from Peru were not resistant to antibiotics, the incident showed how quickly bacteria could spread through international trade. "There are other plausible scenarios on how DT104 got here, such as by dissemination via breeding stock," Angulo said, "but dissemination via fish meal is one way."

"The main point," Angulo concluded, "is that DT104 is a

complex story of animals, their diets, food production, and global commerce. The story has many interlocking pieces, but it comes down to people impacting global systems and disrupting the natural ecology of animals through artificial diets and intense husbandry. This, in turn, impacts our health."

7.

By Wednesday, May 21, Cynthia Hawley had emerged from her stupor long enough to catch the *CBS Evening News* with Dan Rather, who happened to be reporting on another deadly DT104 outbreak in England. The next morning, when Vijups came into her hospital room to report the U.S. Department of Agriculture's test results on her bacterial culture, Hawley interrupted: "It's definitely 104. I saw the news last night. Once the cows' diarrhea started getting watery and bloody, they were dead."

"Yeah, DT104 is what we're dealing with," Vijups confirmed. "The good news is that it's sensitive to fluoroquinolone, the drug you're on."

Although she didn't tell Hawley at the time, Vijups had a deep personal response to Hawley's illness: Vijups' own grandmother had died from salmonella food poisoning decades before. "The stories my mother told about my grandmother's death were always with me when I was treating Cynthia," she said. "I was haunted by the picture of what the world must have been like with no antibiotics to treat the illness. As a physician who encounters treatment failures because of antibiotic resistance, I have moments of fear that we're moving back to that time when infections, even mild ones by current standards, will become fatal again."

8.

Hawley was lucky: fluoroquinolone worked. She was one of millions of beneficiaries of a drug that, when it came on the market in the 1980s after more than thirty years in development, was immediately hailed as a breakthrough treatment for many infections, including the severest cases of salmonella food poisoning. In retrospect, given the drug's unique value in treating potentially fatal human disease, it is amazing that several European countries quickly approved the life-saving antibiotic to prevent outbreaks of diarrhea in calves and respiratory disease in overcrowded poultry—potentially engendering a new round of antibiotic resistance that would put people in even worse shape than before.

Such approval occurred in the Netherlands in 1987, for example. And indeed, soon the bacteria responsible for a major type of human food poisoning began to show resistance to fluoroquinolones. Six years later, in 1993, fluoroquinolones were licensed for veterinary use in Denmark. Soon, the first signs of human resistance to the drug were documented: in 1998, five people associated with a Danish swine slaughterhouse were stricken with a fluoroquinolone-resistant strain of DT104. The bacteria had quickly spread from the pigs to the slaughterhouse workers, who in turn infected nurses at the hospital. The bacteria also contaminated some meat products, infecting a woman who tasted a raw meatball before frying it. In the end, more than twenty additional people fell ill, eleven were hospitalized, and two died.

In 1993, the United Kingdom licensed fluoroquinolone for treating and preventing illness in turkeys and chickens. Two years later, 16 percent of *Salmonella typhimurium* DT104 cultured from farms in the United Kingdom showed some resistance to the drug, and by 1996, fluoroquinolone-resistant salmonella infections were sickening people. Apparently the drug-resistant bacteria had jumped from animals to people, most likely through contaminated food.

Given the rapid development of fluoroquinolone resistance in the Netherlands, Denmark, and the United Kingdom following its use in agriculture, the FDA seemed to have an airtight case for rejecting the drug manufacturer's application, in 1995, to sell fluoroquinolones for use in poultry in the United States. On this subject the FDA also had the full support of the CDC, which sent nine letters to the agency urging it to reject the application. Nevertheless, in 1995 the FDA granted approval for use of the cutting-edge antibiotic to treat respiratory disease in poultry. And sure enough, by 1997 salmonella in the United States had begun to show resistance to fluoroquinolones. By 2000, 1.4 percent of salmonella infections showed some resistance, with the percentage quickly rising.

Another leading cause of food poisoning in the United States, a genus of bacteria called *Campylobacter*, was becoming fluoroquinolone-resistant even more rapidly than was salmonella. In 1997, when researchers from the Minnesota Department of Health tested ninety-one meat products from supermarkets in the Minneapolis–St. Paul area, 14 percent of the bacteria showed fluoroquinolone resistance—up from only 1 or 2 percent in 1992. This was strong evidence, according to the FDA, that the use of fluoroquinolones in poultry posed a risk to human health. Fortunately, this time the agency proposed revoking the license for use of the drug in poultry—with little resistance from Congress. Although Abbott Laboratories, one of the two manufacturers of fluoroquinolones for poultry, withdrew its product even before the FDA officially proposed the ban, the other—Bayer Corporation—reacted far differently, as did many others in the industry. The director of government and industry relations for Bayer, veterinarian Dennis Copeland, insisted, "The consensus is that there is no public health risk."

Alexander S. Mathews, president and chief executive officer of the Animal Health Institute in Washington, D.C., which represents manufacturers of pharmaceuticals, vaccines, and feed additives, claimed that "there is no scientific evidence that salmonella food poisoning has been linked to farm use of antibiotics." Richard Carnevale, also of the Animal Health Institute, declared, "There is no clear documentation that use of antibiotics in these animals was responsible for the emergence of the multi-drug-resistant strain of Salmonella."

Patrick Pilkington, vice president of Live Production

Services at Tyson Foods, stated that "scientific information currently available shows no conclusive evidence of a connection between the veterinary use of fluoroquinolones and antibiotic resistance in humans."

In 2001 Richard L. Lobb, a spokesperson for the National Chicken Council, told a reporter for the *Village Voice* that fluoroquinolone actually "improves the gut health of the bird and its conversion of feed. . . . And if we are what we eat, we're healthier if they're healthier." Of course, the birds themselves, often deformed or weakened by their artificially rapid growth from unnatural feed, their cramped and unsanitary quarters, and their water supply, which is sometimes laced with antibiotics, were profoundly unhealthy. And the spokesperson also neglected to mention that, by eating the birds, people risked ingesting antibiotic-resistant bacteria.

Meanwhile, scientific warnings from the CDC, the FDA, and the American Medical Association largely echoed the warnings from the 1960s that the use of antibiotics for promoting growth in farm animals should be banned. Knowledge had marched on even as common sense stood still.

Antibiotics used on the farm not only make livestock-associated bacteria resistant to some antibiotics but can also remain active after passing through the animals. The drugs then end up in bacteria-rich waste lagoons, and this medicated sludge is often spread on croplands as fertilizer, where the antibiotics and drug-resistant bacteria enter groundwater or surface water and then infiltrate the soil. For example, the

CDC has found significant levels of three different antibiotics in lagoon wastewater drained from industrial feedlots, agricultural drainage wells, and associated water sources. This wastewater contaminates streams, rivers and aquifers, and lakes and their shores, exposing those who swim there or eat fish from the seemingly pristine waters.

Fluoroquinolones have also been detected in wastewater treatment plants in Europe. One study found high levels of antibiotics, very likely from nearby cattle operations, in two lakes in Switzerland. And researchers have reported the presence of antibiotics in river water and sediments in Italy. Other antibiotics have been detected in sediments under fish farms.

If the DT104 outbreak represents the latest epidemic of drug-resistant *Salmonella typhimurium*, it certainly will not be the last. Even the removal of fluoroquinolones from the medicine chests of livestock producers will leave many other antibiotics there. In 1998 the FDA collected 200 samples of chicken, beef, turkey, and pork from three supermarket chains in the greater Washington, D.C., area and found that one in five was contaminated with various strains of salmonella—including DT104. The FDA also identified a strain showing resistance to twelve different antibiotics.

It is only a matter of time, Angulo and others worry, before our brushes with drug-resistant salmonella will lead to a head-on collision with a full-force epidemic capable of killing thousands of people. "It may never happen," he concedes. "But why would we continue to invite such a catastrophe?"

9.

On May 26, 1997, Cynthia Hawley left Northwestern Medical Center and returned home to Heyer Hills Farm. As we sat in the farmhouse kitchen on a blistering July afternoon in 2002, she lamented the globalization of world trade or whatever it was that permitted DT104 to be visited upon her farm and family. Outside the kitchen window, whose sill was adorned with a plaster cast of a black-and-white Holstein, a row of evergreens stood guard along the edge of the north pasture.

Hawley was stunned by Angulo's theory that parts of the bacteria that struck her farm could have come from as far away as Asia. The notion that pieces of a bacterium could hop from fish in Thailand to, perhaps, a bird and then reassemble themselves as they travel across continents, only to strike at the very heart of her family's health and income, was a sobering reminder of the dangerous complexities of modern life. She nodded as if to acknowledge not only the logic of this scenario but also its inevitability. Species intermingle all the time. Humans are connected not only to one another but also to the myriad other species, seen and unseen, with which we share the earth.

"Our family farm, with about 200 head of cattle and 600 acres, used to be considered really huge in this area," Hawley said. "We were a big fish in a small pond; now we're a small fish in a big pond of corporate agriculture. You have to get bigger to survive because you need to produce quantities in order to compete. 'Farm' is becoming a misnomer. It's pretty much industry now. I do not like to see what's happening to the animals because of it."

Later, we walked out to the garden, which Cynthia's mother, Marjorie Heyer, was tending. "The more intensive farming gets, the more props you need," Heyer said. "You crowd the animals to save every cent you can on space; then you have to give them more antibiotics to keep 'em healthy. I'm not saying that's where DT104 came from. I'm just saying that forty years ago what we worried about was nutrition and how to feed the cows right. Now it seems like there is a lot more we have to deal with and worry about, especially after this DT104 thing. What's next?"

Following her parents' inspiration, when Cynthia was twenty-nine she married a farmer, Brian Hawley, who owned a large independent dairy operation nearby. Even then, it wasn't as if the couple had a secure hold on their dream. "One January morning my husband went out to start a tractor. The tractor ran over him, and he was killed," she said, her stare fixed on a distant memory. Determined not to let their vision die with him, Cynthia operated the large farm on her own, with hired labor, for the next twelve years. In the fall of 1996 she sold the farm and moved back to Heyer Hills Farm, where she had grown up. It was not exactly as if she had come full circle; she had traveled down a river. Life had always been a river, but now, fed by new tributaries from all parts of the world, the river of life at Heyer Hills Farm felt swifter, more dangerous, and less predictable than ever before.

Of Old Growth and Arthritis:
Lyme Disease

1.

About the year 1700, John Harrison of Long Island, New York, bought from the Lenni-Lenape Indians a 17,000-acre tract of oak-hickory forest near what is today the city of New Brunswick, New Jersey. At the time, Harrison's purchase was but a tiny grove within 100 million acres of woodland that stretched from Virginia to New England and west to the Mississippi River.

In 1701 Harrison sold 10,000 acres of his land to a group of Dutchmen, who divided it into eight parcels. South Middlebush, the first road through the region, crossed the tract north to south, subdividing the eight parcels into sixteen, and other subdivisions began to be made. Cornelius Wyckoff, one of the buyers of Harrison's land, for example, gave 300 acres to each of his four sons, who built houses and cleared land for crops and livestock, leaving behind several forested woodlots for timber and firewood. By the mid-1800s farmland quilted the region, a railroad had arrived, Indian footpaths had become roads for horse-drawn wagons, and proliferating byways had further fragmented the remaining forest lots. Old Indian Path, the easternmost boundary of the original Harrison land, soon became the busy Lincoln Highway, which carried automobiles between Philadelphia and New York City.

Changes in this part of the country mirrored what was happening to forests throughout many settled regions of the Northeast. Farms were built, forests cut, and by 1800 the 4 million settlers in the Northeast had spilled into the remotest corners of New England. At farming's peak around 1900, more than half of the 100 million acres of northeastern forests had been cut, including all but a fraction of Harrison's original land. By that time the largest stand of original trees in his 27-square-mile purchase was a 65-acre woodlot. That forested enclave must have been a spectacle even in the mid-1800s: two- and three-hundred-year-old trees presiding over a shadowed realm of birdsong, butterflies and flying squirrels, grouse, turkeys, and perhaps a bear or panther passing through on its way inland or farther north.

In the mid-1950s, this final trace of the original forest came under assault from a timber company hoping to liquidate the valuable hardwood. To protect this rare jewel of nature, in 1955 several organizations, including the United

Brotherhood of Carpenters and Joiners of America, purchased the land, named it in honor of a former union president, and donated it to nearby Rutgers University. Today, hidden between sprawling New Brunswick and Somerville, New Jersey, the 65-acre William L. Hutcheson Memorial Forest remains one of the largest old-growth oak-hickory forests in the mid-Atlantic states.

2.

On a hot morning in June 2001, three hundred years after Harrison's purchase, I drove to Hutcheson Memorial Forest. There I was greeted by the forest's director, Edmund W. Stiles, a Rutgers ecology professor. After we introduced ourselves, we walked down the path into the ancient grove.

The air turned cooler as the bright morning light dissolved into the soft hues of the forest edge. There seemed to be as many fallen trees as standing ones. Although not the massive, moss-draped druids of the purple prose often used to describe a primeval forest, they were the biggest trees I had ever seen in New Jersey. Their massive branches created a heavy latticework against the blue sky.

"Ecologists once believed that forests reached a climax and would stay that way," Stiles began. "Maybe that's where the notion of the forest primeval arose. But it's not like that. About 250 years is the age of old trees in this patch. The oldest tree ever recorded here was 344 years old, from 1611. It was blown down in a hurricane in 1955."

Stiles said that since the forest's beginning—sometime after the last ice age, 10,000 years ago—natural catastrophes, especially epic storms, had struck every few centuries. Truly ancient trees aren't seen there because they get knocked down. Put another way, the forest is ancient, but the trees are not.

"One way to talk about an old-growth patch like Hutcheson isn't in terms of the age of the trees but in the length of intervals between major natural disruptions. It's not just about the trees but about the process, the whole system. Another way to think of an old-growth forest is as a place where trees die natural deaths rather than getting cut down."

Is, then, a healthy forest as much about dead trees as live ones, I asked. Stiles nodded and explained that a dead tree provides an opportunity for numerous insects, birds, and mammals to contribute to the forest for hundreds of years. When a tree falls, mosses and other plants colonize it. Even the hole left in the ground by upturned roots becomes new habitat for small, enterprising species. But it takes a long time for trees to die and begin to return to the soil—a luxury of time that many of today's forests don't have. "It disappoints me when forest managers talk about having to clean out 'dead wood,'" Stiles said. "In cleaning out dead trees you destroy habitat that makes a healthy forest."

A severely disrupted forest can quickly lose many of its most "specialized" species—animals that can't quickly adapt to new habitats or sources of food—Stiles continued. "Generalists," on the other hand, often accommodate change. While specialized species vacate the forest, the resourceful generalists, such as deer and mice, often expand their numbers.

"The age of trees is a huge influence on the animal community, and one of the many ways eastern forests have been degraded is by keeping them young. Another way is by fragmenting a forest into patches, like here at Hutcheson. If a forest is under a certain acreage, many animals can't live there."

There is almost as much forest in the East today as there was two centuries ago, but its pattern is quite different now, Stiles went on. After peaking about 1900, eastern farms declined as western trade routes opened markets to cheaper midwestern grain. Where the farms were abandoned, trees often grew again. By the early 1900s forests had returned to cover 50 million acres, and today forests cover three-quarters of their historical range. But one should not be fooled by size alone: even though the trees may have returned, the forests have not. Farms flowed in with people and livestock and then washed out, taking with them mountain lions, wolves, bison, wolverines, elk, mountain lions, bobcats, fishers, and numerous other species.

As we continued our stroll, I commented that the forest interior was browner than I would have imagined for an ancient forest—far from the deep, leafy tunnels, green boughs, and verdant undergrowth I expected to see. Stiles pointed out that the thick canopy of the dominant oaks, hickories, and, especially, sugar maples filters light.

"Availability of sunlight influences what grows and doesn't grow on the forest floor," he continued, explaining why some old-growth forests can be remarkably park-like in effect.

A fragmented forest receives more sunlight than a contiguous one because every road, clearing, or cut is a virtual skylight. The light fosters more leafy growth along the forest edge, which provides browse for deer. That's one reason why deer do so well in fragmented forests. Where there's sunlight, there's browse, and browse attracts deer.

"Deer thrive and forest sickens," I commented, paraphrasing a recent headline in New Jersey's *Star-Ledger*. "The article said that white-tailed deer were almost extirpated from New Jersey a hundred years ago. But now they number in the hundreds of thousands."

Stiles snorted. "Those deer are living on borrowed time," he said. "Three hundred years ago there were probably about 25 per square mile. Now there are something like 200 per square mile. Something will knock them down, and it could be disease."

He paused on the trail and placed his palm against the bleached skeleton of a dead oak. As if to highlight the contrast with other forests in the region, Stiles explained that the dead tree had been standing when he first came to this patch, more than thirty years ago. When the next generation of biologists comes to this forest, the toppled trunk will, perhaps, have disintegrated and become little more than a raised ridge in the ground. Yet it will still be contributing to the forest's health.

Stiles pointed toward the forest edge as we continued on. Through the brush could be seen a distant cultivated hillside and a row of houses on the horizon. I pressed the soles of my shoes hard against the never-tilled earth. The roots below clutched soil and boulders made from glaciers that had retreated 10,000 years ago and the stone arrowheads of ancient hunters who had passed through. As I gazed upward, the outstretched boughs of an oak seemed to embrace and hold me. What threads we silently break; what voices we still. By what grace, I wondered, have we been kept so well by what we have abused for so long.

3.

Like much of the Raritan River valley, Hunterdon County, New Jersey, has seen some of the most rapid development in the East. Drained by three graceful rivers, cloaked by beautiful, if young, forests, and within commuting distance of New York City, Hunterdon has been transformed over several decades from countryside to a suburbanized hub with more than 120,000 people.

"When I got here in 1985, I thought I was coming to a quiet rural county," said John Beckley, who lives in the town of Annandale, about two miles from a branch of the Raritan River. As the county's director of public health, Beckley has seen firsthand many of the changes wrought by population growth and commercial development. "In one respect we're no different from many other places in America. It's just happening a lot faster here," he said. "When I got here I expected to deal with bread-and-butter public health issues. We had an environmental staff of four and spent most of our time inspecting or issuing permits for septic systems and wells—a dozen or so a week. Since then, my job's gotten a lot more interesting."

In 1985, only months after Beckley arrived at his new job, New Jersey reported its first case of infection with the human immunodeficiency virus (HIV), which had been isolated and described two years before. In 1986 the county had its first cases of another new illness, Legionnaire's disease. Two custodial workers, who survived, were stricken at Hunterdon Central Regional High School, Beckley said. Eventually, the Centers for Disease Control and Prevention discovered that bacteria living in a poorly designed water heater had caused that particular outbreak.

Next, in 1989, Beckley's office began receiving reports of raccoons behaving strangely, wandering across the county's highways and into people's yards to attack their dogs. "It turned out to be the first outbreak of terrestrial rabies in the state in nearly half a century," he said.

Ten years later West Nile virus arrived, an event that led to the county's first mosquito control program. "We have a whole lab now, several trucks, a special freezer for preserving specimens at minus 70 degrees, and a mosquito-control team," Beckley said. "You can imagine how our staff and budget have grown since I got here in 1985. None of the traditional public health challenges have gone away. The fact is, we have more infectious diseases than before."

Many of the diseases that suddenly struck Hunterdon County were not random: they were precipitated or fostered by human changes to the environment. The rabies outbreak, for example, was traced to hunters who transported raccoons from Florida and released them locally to improve hunting farther north, in West Virginia. Some of these raccoons were infected with the rabies virus. From there the virus marched north through the species, right into New Jersey. Legionnaire's disease, a technology-related illness, was caused when the ubiquitous *Legionella* bacteria were given the opportunity to collect in warm environments provided by modern life, such as water heaters, saunas, and air conditioners, and were then aerosolized and inhaled by people nearby.

"All these diseases were occurring against the backdrop of what has become our single biggest infectious disease problem," Beckley continued. He was referring to Lyme disease, the most common vector-borne illness in the United States. It's another disease that accumulating evidence indicates has emerged in part because of radical changes people have made to the landscape—in this case the once comparatively stable and biologically rich forests of the eastern United States, of which Hutcheson Memorial Forest is now only a sad token.

According to Sarah E. Randolph, a professor of parasite ecology at the University of Oxford, England, "It isn't known when the Lyme disease bacterium was first introduced into the United States, but it is difficult to believe that it is as recent as the last major resurgence." In other words, the bacterium that causes Lyme disease very likely has been in the United States for a long time, but until recently the conditions did not exist for the disease to become epidemic. Randolph said that though "no one can say for sure when the disease first appeared in the U.K., the bacterium has been around for a long time, at least in Europe."

In the United States, Lyme disease was first described in Old Lyme, Connecticut, in the 1970s, but a case wasn't documented in Hunterdon until 1988, when twelve patients were identified. There were 30 cases in 1989, and by 1993 there were 204. Today Hunterdon County has the third highest rate of Lyme disease in the United States and the highest in New Jersey. In 2000 Hunterdon had more than 500 cases. Only Nantucket, Massachusetts, and Columbia County, New York, have higher rates, according to Beckley.

The CDC sent a team to investigate—Beckley was one of the members—and concluded that one reason for the high incidence in Hunterdon was the county's high density of deer, which harbor the tick that carries the Lyme disease bacterium, near residential areas. Another was the high number of rock walls and woodpiles near homes. These provide refuge and breeding grounds for mice and chipmunks, which, it turns out, also carry the ticks. "Where the edge of a yard comes up against the woods, that's an 'ecotonal edge,' which is perfect habitat for ticks," Beckley explained. "That nature-culture border is where people, who may be mowing the lawn or trimming branches, often pick them up. Human activity has put people right at the center of the tick's life cycle."

Deer and small rodents such as mice are the literal lifeblood of the ticks. These mammals provide not only blood meals but also a means of transportation and dissemination for the otherwise largely immobile ticks. The life cycle begins in fall, when the egg-laden females drop from the deer to the ground, frequently nestling in leaf litter for the winter. With the advent of warm spring weather, the eggs hatch and the larvae hitch a ride on mice, chipmunks, or any other small mammal or bird nearby. Once on a host, the ticks feed for several days and then drop off. They develop over the next several months and re-emerge as nymphs the following spring. By then they have become mobile enough to climb low-lying bushes, where they often perch at the end of a branch or leaf and wait for a larger mammal, such as a deer, to pass by. A horse, dog, or human will do. It is by these poppy-seed-size nymphs that most people become infected.

Hunterdon County apparently didn't even have deer ticks until the mid-1980s, according to Beckley. Or at least not enough to notice. An increase in deer numbers and perhaps a warming climate may have increased their numbers. Over the past century, the average temperature in nearby New Brunswick increased by almost two degrees and precipitation increased in that part of the state. These climatic changes created ideal conditions for the ticks. The changes also generally paralleled an explosion in tick populations throughout the northeastern and upper north-central United

States. In Hunterdon County, ticks found an estimated 30,000 deer to feed on—more than in any other county in the state. "Hunterdon County may be God's country, but it's also tick country. At least now," Beckley quipped.

There was some hope that hunting would reduce the deer population. "But most hunters want to shoot antlered bucks," Beckley explained. "Because the bucks are polygamous, even if their numbers are reduced markedly, most of the remaining females will still likely get pregnant, ensuring a high birth rate the following year." In an effort to tip the balance, officials of the New Jersey Division of Fish and Wildlife initiated an "Earn a Buck" program whereby a hunter who kills an antlerless deer—presumably a doe—can then legally shoot a buck.

Rapid development is quickly neutralizing the potential benefits of hunting in reducing deer, however: it's illegal to hunt within 450 feet of a residence without the owner's permission, and many open tracts of land are privately held and not accessible to hunters. The new houses going up in the county are therefore creating more safe havens for deer. All the environmental, economic, social, and political dynamics are thus weighted in favor of deer herds living in proximity to humans, and this increases the risk of people getting Lyme disease and other tick-borne illnesses.

Experts have considered other approaches to reducing the deer population, including netting herds and then humanely killing them, using sharpshooters, or even instituting birth control. But these are expensive or unproven solutions, and discussion of them frequently arouses the concerns of animal rights proponents. "Politically, bringing about a significant reduction in the deer population is a very difficult goal to accomplish," Beckley said.

In concept, Lyme disease should be easily preventable. Prevent tick bites and you prevent the disease. For many years, the Hunterdon County Department of Health has had Lyme disease education and awareness programs in place. Several full-time staff members, including a health educator, are at work on these efforts. "Despite stressing to the public the importance of prevention—use insect repellant, avoid tick areas, stay on trails during tick season, wear light-colored clothing, and check yourself carefully when you come inside—we still can't seem to decrease the county's infection rate," Beckley said.

4.

"In early October 2001, John and I were driving home from a weekend on Cape Cod and I started feeling this really pronounced stiffness in my spine," Linda began. "All my muscles hurt. I lost my appetite and got very agitated. I got a fever, and shivers came in spasms." The night after a nurse practitioner diagnosed her illness as flu, John Beckley noticed the telltale bull's-eye rash on his wife's right shoulder blade. She had Lyme disease, he felt certain. A visit with her doctor and a three-week course of antibiotics cured her symptoms. She was lucky to have been quickly diagnosed; many people don't realize they have Lyme disease until the symptoms are far worse, and in some cases permanent, including painful joint or neurological damage.

Linda Beckley told me she wasn't sure where she picked up the tick, but she believed it happened as she was walking their dog, Willie, near the South Branch of the Raritan, not far from home. Twice before, she had found ticks in her car after such walks.

We walked outside to the back porch. As with many homes in the community, the Beckleys' backyard is carved out of the woods. Earlier, I had driven through this urban archipelago, passing islands of lawn in a sea of fragmented forest. Wooded peninsulas wrapped behind houses and extended into front yards. Paved roads ran around the neighborhood in the maze-like geometry of an integrated circuit board. Grass abutted woodlands everywhere, and the deer, squirrels, and chipmunks I saw that afternoon readily crossed between both. Ornamental shrubbery lined the foundation at the front and back of the Beckleys' house. It was dream habitat for deer.

Linda and I drove toward the park to walk Willie. "When I moved here three years ago, all this used to be a big farm," she said, sweeping her hand above the steering wheel as we left their neighborhood and entered a new housing development. "When John got here, beyond the farm was all woods. Now it's all these new houses."

With the houses came legions of people suddenly thrust within arm's length of the deer that came to feed on the lawns. Along new clearings, leafy browse flourished. Rock walls were built at the perimeters of properties, and wood-piles appeared at the edges of driveways, creating a paradise for the carriers of Lyme disease ticks.

When we got to the park, Linda opened her door, and before she had even unrolled the leash, Willie barreled out and romped across a field toward the river. As we walked toward him we passed a wooded area, catching a glimpse of five deer in chocolate brown coats. Willie, meanwhile, had swum across the South Branch and was heading up an embankment beneath towering cement pylons supporting Interstate 78, where four lanes of traffic roared above the opposite bank.

"Get back here, Willie!" Linda called. While we waited, I bent down, picked up a dried gray branch, and tossed it into the river. The current grabbed it and swirled it around. Willie finally returned and scrambled up the wooded stream bank toward us.

"He does this sometimes," Linda said apologetically.

On the drive back that evening, I was delayed by the aftermath of a traffic accident on I-78. As I waited at a stand-still, I imagined how far along its journey toward Hutcheson Memorial Forest the stick was that I had thrown into the South Branch of the Raritan. And that brought me back to the June morning several months before when I spent time in the old-growth forest with Edmund Stiles. Age is defined not only by objects such as trees but also by the subtle processes of a growing forest, I thought. Where, I wondered, were the long, uninterrupted interludes of our world—a world in which intervals between major changes seem to shrink ever smaller day by day: neighbors moving to new jobs in other cities, a mall newly constructed here, a farm giving way to a new housing development there.

Even the intervals of Linda's disease had been shrunk, in a sense. Her doctors had compressed the definition of her Lyme disease into the interval between when she was bitten and when she successfully completed her course of antibiotics. But this clinical definition excluded the larger ecological implications of her illness and therefore its full meaning. Her illness was not just about a bacterium that entered her body. It was an extension of the unfortunate history of the eastern forests, and it was connected to autumn oaks and hickories, an absence of predators, and an overabundance of deer and mice. Her illness was not exclusively hers. It was an intimate part of a picture almost too big to see.

5.

Perhaps no one understands this big picture—the ecology of Lyme disease—better than Richard Ostfeld, an ecologist with the Institute of Ecosystem Studies in Millbrook, New York. In the mid-1990s, Ostfeld began to suspect he might predict people's risk of contracting Lyme disease based upon, of all things, the abundance of acorns in a region. Acorns come in bursts, or "masts," with almost none produced in some years and bumper crops produced in others. These cycles are synchronized among trees over large regions of the country, in part by regional weather.

If acorns attract deer and mice, Ostfeld reasoned, and the incidence of Lyme disease in humans is related to the densities of these animals, the rate of human infection could be related to the production of acorns. Ecologists call it a cascade effect.

The year 1995, a very poor one for acorn production near Millbrook, gave Ostfeld and his colleagues an opportunity to test his theory. Millbrook is in Dutchess County, which has an unusually high rate of Lyme disease. With lots of oaks and people, it was an ideal place to conduct the study. Ostfeld and his team measured and demarcated two sets of plots in the forest at the institute. On half of the plots, Ostfeld let the poor natural acorn crop fall. On the other plots, he supplemented nature's production with nearly a million acorns from elsewhere. In the months that followed, he regularly visited the plots and compared what happened in the supplemented plots with developments in the acorn-poor ones.

For one thing, he noted, the supplemented plots attracted far more deer that fall. The following spring, something else was evident: mouse populations had exploded in these same plots because more of the well-fed adult mice survived the winter than had their poorer cousins, and they had more young in the spring. Ostfeld and his colleagues dragged strips of fabric over the plots, a standard method for collecting ticks. Astonishingly, the acorn-rich plots had eight times as many newly hatched ticks, or larvae, as the regular plots. The acorns had not attracted the ticks. Rather, the greater number of deer attracted by the abundance of acorns meant that more adult ticks had dropped from the deer as the animals fed that fall. In the spring, the female ticks on the ground laid eggs, which hatched into a superabundance of larvae by early summer of 1996.

There was something remarkable about these newly hatched ticks: they didn't harbor the bacteria that cause Lyme disease. Even if the mother tick was infected, the bacteria were not passed through the eggs to the larvae. To become infected, a tick first had to feed on an infected animal, such as a mouse. And since almost all mice carry the bacteria, almost every tick that feeds on a mouse becomes infected.

It stood to reason that the more mice there were in an area, the more likely it was that actively feeding ticks in that area would become infected. And since more mice had been drawn to the acorn-rich plots, Ostfeld was not surprised to find a higher percentage of infected ticks there. Moreover, since a higher density of infected ticks led to more Lyme disease in people, Ostfeld's acorn theory was supported by these field experiments. Acorns attract deer and mice, mice infect ticks, and infected ticks give people Lyme disease. People's health was linked to acorn production.

The real test of Ostfeld's theory would occur after a natural mast. He could then compare human infection rates in regions where a bumper crop occurred with infection rates in places where it had not, or he could compare infection rates in the same area over several successive years. A spike in infections at a certain interval after each large acorn crop would lend considerable weight to his theory.

6.

The year 1997 saw one of the most prolific acorn crops in the mid-Atlantic states in years. In Hutcheson Memorial Forest, Edmund Stiles recalled the abundance of acorns that year. Moving through the forest, he said, was "like walking on marbles." Perhaps no place experienced a greater rain of the acorns than the beautiful, oak-arched campus of Drew University in Madison, New Jersey. The campus is situated in a grove of old oaks, many of them towering a hundred feet overhead. The school has been called the "university in the forest."

Twenty-year-old Jeff Dunbar was a sophomore in the fall of 1997, and so many acorns pummeled his dorm roof that term that the noise sometimes woke him at night. By day, deer from a nearby forest wandered across campus, and chipmunks, squirrels, and mice scurried amid the landscaping at the building's foundations. The presence of so many deer on campus that fall meant that numerous egg-laden ticks were falling to the ground. And two years later—by the summer of 1999—these led to a population explosion of blood-hungry nymphs. That happened to be the very summer Dunbar decided to live on campus while working for a state assemblyman nearby. Dunbar played Frisbee on campus almost every evening, and wayward throws often sent him scrambling through tick-laden bushes.

On August 1, Dunbar awoke with the left side of his face paralyzed. He went to the emergency room at nearby Morristown Memorial Hospital, where the attending physician treated him with steroids. A Lyme test, which measures the body's immune response to the Lyme disease bacterium, was negative. But the test can be inconclusive if the body's immune response has not yet kicked in. The disease can therefore go undetected. His facial paralysis disappeared two weeks later without further treatment.

He continued his job into the winter of 2000. That February, as he was stuffing mailings into envelopes, his shoulders and elbows became so stiff and sore that he could hardly move them. His physician concluded that the repetitive stuffing motion had strained both shoulder joints, and he recommended physical therapy. Dunbar improved, but he still felt tired and sore. During the summer of 2001, as he was undergoing routine arthroscopic surgery for a slight tear in a knee ligament, his surgeon discovered severe inflammation in the joint. A Lyme test ordered by the surgeon was positive. By now, the bacteria had invaded Dunbar's joints and spinal fluid. After eight weeks of intravenous antibiotics, he improved, and his symptoms largely disappeared.

Meanwhile, Ostfeld had gathered data on the rate of Lyme disease in the mid-Atlantic states, where a large acorn mast had occurred two years earlier. If his theory was correct, the rate of infection should rise among people there in the second year after the mast. The 1999 infection rate did increase in the area affected by the mast of 1997. In fact,

1999, the year Dunbar became ill, saw the third-highest number of Lyme disease cases ever reported in the mid-Atlantic region.

7.

Mice and chipmunks transmit Lyme disease to more than 90 percent of ticks that feed on them, whereas possums, raccoons, birds, and many other forest dwellers infect only about 10 percent of their ticks. This contrast goes to the heart of the ecology of Lyme disease. The feeding options for ticks increase in almost direct proportion to the variety of species from which they can choose. If a forest held a greater variety of animals—say, as Hutcheson Memorial Forest once did—the probability of a tick feeding on a mouse would be reduced. And the chance of the tick picking up the Lyme disease bacterium would thus be reduced.

Ostfeld wondered if this meant that the loss of species in the northeastern forests, which now favored generalists such as mice and chipmunks, contributed to the increase in Lyme disease. Conversely, if a greater variety of species were returned to a forest, would that reduce the density of mice and chipmunks and therefore the high rate of infection among ticks—and people? Would a greater degree of biological diversity, in other words, offer people a certain degree of protection from Lyme disease?

Ostfeld couldn't recreate the rich and diverse forests of old, but he could test his so-called dilution theory with computer modeling. So he created a computerized forest. Each time he added a new species to the computer-modeled forest, the density of ticks infected with the Lyme disease bacterium declined.

Ostfeld and his colleagues tried to get a real-world grasp of his hypothesis by listing all the eastern bird, mammal, and lizard species, from Florida to Maine, on which ticks carrying Lyme disease are known to feed. The farther south one moves, the greater is the diversity of species. The researchers then compared the numbers of different species within regions along the eastern seaboard with the rates of Lyme disease in people in those same regions. They found that the areas with more species had fewer cases of Lyme disease per capita. High biological diversity, it seemed, did tend to minimize the rate of Lyme disease infection in the human population—at least that was one reasonable interpretation of his findings.

In considering how to lower the risks of Lyme disease, many more ecological questions need to be answered. Ostfeld is now trying to understand, for example, the effects of habitat fragmentation on Lyme disease risk for people living nearby. What animal species are our strongest allies in protecting us from Lyme disease? And just how big need a forest patch be to support many species and, therefore, lower the risk of the disease?

8.

The ecology of Lyme disease reminds us that the connections between the earth and human health are ancient.

Changes in forests and their species, research such as Ostfeld's suggests, are reflected in human disease. Places such as Hutcheson Memorial Forest are touchstones of seeming tranquility in a world undergoing constant change by humans. We will probably never know if Lyme disease afflicted forest dwellers there five hundred years ago, but the diverse ecology at that time would have weighed against it. If the Lyme disease bacterium were present, the indigenous forest dwellers might, over millennia, have developed immunity to it. What we can be sure of is that, in our shortsighted efforts to make the world more hospitable for humans, we have been making it more hospitable for many of the microbes that cause disease.

...5...

A Spring to Die For: Hantavirus

1.

The mysterious illness that killed two young Navajos in the spring of 1993—and the deaths that followed—captured the nation's attention like no other outbreak since Legionnaire's disease in 1976. The victims, Merrill Bahe, 20, and his fiancee, Florena Woody, 21, were young and had no medical history that might explain why they had become ill. The doctors who treated them had never seen anything like it.

Merrill Bahe and Florena Woody grew up in two starkly different worlds on the same Indian reservation, 25,000 square miles of land extending into New Mexico, Arizona and Utah.

For Merrill Bahe, each day brought a new struggle with poverty. Each weekday morning, Merrill would awaken at 5 a.m., slip out of his blanket—the only bedding he had ever known—and creep toward the door of his family's wood-and-tar-paper shanty.

It was as much to ease the burden on his family as to get a square meal that Merrill Bahe would time his hour-long run so that he would arrive at Torreon Middle School two hours before his first class, in time to eat breakfast with the kindly kitchen staff. One day, Torreon's track coach, impressed with the boy's speed and strength of character, called track coach Mike Gorospe at the Santa Fe Indian School, a boarding school set up by the U.S. government and now run by the 19 Pueblo tribes.

Merrill Bahe's acceptance to boarding school for the fall term in 1988 promised to change his life.

Florena Woody had always been lively and enthusias-tic, eager to ride spirited horses and climb the cliffs behind the family's trailer encampment in Littlewater, N.M., to pil-fer feathers from an eagle's nest. Now she had a new radi-ance. Florena's brother Collins remembers that every time the talk turned to Merrill Bahe, Florena Woody "just lit up."

Florena Woody's illness began on April 29, 1993, with nothing more alarming than muscle aches in her neck and shoulders. Four days later, fever set in. She began to cough. On May 6, Florena visited her doctor, who told her it was probably a mild case of the flu. The doctor gave Florena a shot and antibiotics. Nothing seemed to help. One week later, Florena was still feverish. That Saturday night it became clear to Bita Begay that her daughter was becoming increasingly lethargic. She decided to take her to Crown-point Hospital, seven miles away.

"I can't breathe," Florena Woody lamented, as the attending physician examined her.

· · · · · · · ·

The doctor ordered a chest X-ray. Soft tissue normally appears black on X-rays. [The film] gave him a "sinking feeling." Florena Woody's lungs were white. Her lungs were rapidly, and inexplicably, filling with fluid.

From their vantage point in the hallway, the Woody family watched in disbelief as the hospital staff replaced the second bed in the room with a mechanical ventilator. But the task was futile. An alarm shrilled. Faces turned toward the heart monitor. A glowing green wave form told them that Florena's heart had stalled, then stopped.

Merrill Bahe's symptoms, mild at first, worsened two days later. It was Tuesday, May 11. Florena's funeral was just three days away.

"Go to the hospital," Bita Begay ordered. Merrill climbed into the pickup, and Florena's cousin Karoline drove him to the Crownpoint Hospital, where a puzzled young doctor found abnormalities in Merrill's blood test but nothing that revealed what was wrong. The doctors decided to discharge Merrill with stern instructions to return if his symptoms worsened.

By Friday, the morning of the funeral, Merrill's lips were turning faintly blue from lack of oxygen. Collins asked his cousin Karoline to drive Merrill to the hospital in Gallup.

About 10 miles north of Thoreau, Merrill began visibly struggling for every breath. His skin was sallow, and his lips turned a deeper blue. Karoline pulled into the parking lot of B.J.'s Convenience Store, in the tiny roadside community of Thoreau. Paramedics from the Thoreau Volunteer

Ambulance Co. arrived moments later, but they could not revive him.

Patricia McFeeley, deputy director of the OMI, conducted a limited autopsy on Merrill, removing just enough tissue for the state laboratory to test for pneumonic plague, a flea-borne disease that occurs regularly in the Four Corners. But the plague tests, completed after midnight, were negative. Something else had killed the young couple. And McFeeley had no idea what it could be.

—Steve Sternberg, "An Outbreak of Pain"

2.

The Colorado Plateau, which stretches across 130,000 square miles of southeastern Utah, northern Arizona, northwestern New Mexico, and western Colorado, has seen more rain and snow during the last 25 years than at any other time during the past two hundred. And those two centuries have been the wettest on the Plateau in the last 2,129 years. The Plateau is actually a huge basin filled with tablelands and surrounded by mountains. It is a world apart from the rest of the Southwest, older, with its own assemblage of plants and animals and climatic patterns.

The Plateau's remarkable climatic history is told by ancient trees or their enduring remains—some more than a thousand years old. Each spring, beginning at a time lost to all but the memory of these relics, a new growth layer swelled beneath their bark. Toward the end of the growing season, as sap drained away, the layer remained, and the following spring a new one grew. Year by year, century by century, the process continued, creating a trunk of concentric growth rings. Rainy growing seasons tended to produce wide rings, and droughts created narrower ones. By extracting a straw-size core of wood and examining it under a microscope, tree-ring specialists such as Henri Grissino-Mayer at the University of Tennessee, Knoxville, can not only estimate a tree's age by counting the rings but also get a sense of the climatic patterns of its time by analyzing the rings' character and width.

There has been a dramatic long-term climatic shift on the Colorado Plateau, from desert-like conditions to, in more recent times, almost seasonal monsoons. Many climatologists attribute part of that shift to more frequent rises in ocean temperature near the western coast of South America, a phenomenon known as El Niño.

During El Niño years the surface waters off the west coast of South America become unusually warm, whereas during so-called La Niña years surface temperatures cool. These fluctuating ocean-surface temperatures can affect many aspects of the weather by influencing the amount of water that evaporates into the atmosphere and the course of high-altitude winds. The shift between El Niño and La Niña has historically been transient and mild. But in recent decades El Niño has been unusually persistent, leading to greater extremes in the weather including increased precipitation in many places.

Because El Niño is accompanied by a slackening of the Pacific trade winds or even a reversal in their east-to-west direction, storms that normally pass over the Northwest can shift southward, dumping unusually heavy rains or snows on southern California and the Southwest.

If El Niño itself is natural, the extremes and duration of the heated Pacific appear to be something new—made worse by a warmer global climate, some scientists argue. And that, much evidence suggests, is the result of the present scale and character of human activity—the extent of automobile and truck exhaust, coal-powered generating plants, and other sources that emit heat-trapping gases into the atmosphere. According to Grissino-Mayer, who has studied the historical rainfall patterns on the Plateau, "Global warming is intensifying many of the natural cycles such as El Niño. There's no doubt about it, in my opinion. The two-hundred-year period of increased rainfall also coincides with the increasing use of fossil fuels and the emission of greenhouse gases into the atmosphere."

A powerful El Niño cycle began in 1991, and early in the following year a severe El Niño–driven flood moved across the Los Angeles area, stranding fifty motorists in quickly rising waters and sweeping a fifteen-year-old to his death. The following month, winds laden with Pacific moisture pummeled Las Vegas with two and a half inches of rain, flooding streets and turning Duck Creek into a torrent. In December 1992, unusually heavy snows fell at Gallup, New Mexico, and over much of the Colorado Plateau, transforming the watercolor landscape into a monochrome photograph. Following heavy rains in the area, the Federal Emergency Management Agency declared the normally arid state a flood disaster area. During the first three months of 1993, a series of mild snowstorms interspersed with rain fell across the Plateau, and New Mexico was again declared a flood disaster area, along with neighboring Arizona.

Autumn and winter precipitation increased the soil moisture that fed the juniper and piñon woodlands of the Plateau, helping the piñons produce a huge crop of nuts in the fall of 1993. These were consumed by people, birds, and numerous rodents. Awakened by the fall rains, millions of downy chess grass seeds, scattered the preceding autumn, began to germinate, spreading their roots beneath the moist soil. When the first rains of spring arrived, the extensive roots quickly soaked up the water and gave birth to bright green seedlings with hairy leaves.

Snakeweed also burst forth, creating refuge and food for grasshoppers. This unusual bounty of energy-rich wildflowers, nuts, juniper berries, and cones led to an increased number of mice, whose reproductive cycles were triggered by their consumption of the abundant green vegetation. The heavy rains had changed the cycle of life on the Plateau.

3.

Hantavirus pulmonary syndrome (HPS), a usually fatal infection that causes victims to drown in their own fluids, was not exactly new—or at least not to the Navajo. It was caused, the elders said, by Na'ats'oosi, the mouse, and by ch'osh doo yit'iinii, a tiny invisible presence in the mouse's urine. It entered the nose and mouth and took the victim's breath away. An abundance of mice, Navajo elders said, brings the disease to the Plateau and kills healthy young Navajos. It had done so twice before, they said: once in the spring of 1919, a year after a devastating influenza epidemic struck the Navajo reservation, and again in 1933 and 1934, following unusually heavy winter and spring rains. The elders suspected that after the winter and spring rains of 1992 and 1993, the disease had returned. Florena Woody and Merrill Bahe were two of its victims.

Although the Navajo explanation for the disease apparently goes back for generations, the disease eluded detection by public health authorities until the severe 1993 outbreak was triggered, at least in part, by unusually heavy El Niño rains. Here was an illness rising and falling with rainfall patterns that humans themselves seemed to be influencing, and a haunting example of how the fates of a young couple in Stillwater, New Mexico, were influenced—if not sealed—by fluctuating ocean temperatures off the coast of Peru, thousands of miles away.

As the mysteries of the disease began to be unraveled, understanding of it grew as old Navajo wisdom blended with scientific analysis in a most unusual way. What emerged, at least to those who could hold both perspectives, was a powerfully new, encompassing view of humans not as a stand-alone species but as just one species among many in a web of climate, ecology, and intertwined fates. It was a view of a human illness whose significance transcended emergency rooms and the search for a cure.

4.

One modern interpreter of Navajo medical beliefs is Ben Muneta, a physician with the Indian Health Service in Albuquerque, New Mexico. Born on the Navajo reservation and trained at Stanford University School of Medicine, Muneta was working with the Indian Health Service in the spring of 1993, when a number of people in the Four Corners region were stricken with hantavirus.

"Most everyone seemed to be totally baffled by what this killer disease was, including the Centers for Disease Control and Prevention, which sent a team to investigate," Muneta told me in 2002. In June, Peterson Zah, the president of the Navajo, convened a meeting in Window Rock, Arizona, of Navajo healers to seek their guidance. Each healer spoke about how humans are not the dominant force in nature but instead are dependent upon other forms of life for existence. The outbreak had resulted from disharmony in the environment, they claimed, and now ceremonies were needed to reestablish harmony between patients and the universe.

"People at the meeting, including a few from the CDC, began to realize that hantavirus was not a new disease," Muneta said. "A few elders also spoke of Na'ats'oosi not simply as a mouse but as a 'thing that sucks on things and leaves a trail of saliva as it flees.' Perhaps they realized the virus might be spread by mouse saliva as well as urine. Elders have long had taboos to prevent human contact with mice."

Muneta believes that the Navajo not only had understood the basic ecology of the disease for centuries but also had designed a healing ceremony specifically for it. His belief is based on a Navajo sandpainting, which he photographed, that depicts a mouse and several medicinal plants, two of which went by the Navajo names Tl'oh azihii libáhígíí and awe'e'tsa'a'l.

Tl'oh azihii libáhígíí is a member of a group of plants known as ephedra, Muneta explained. Plants from this group contain ephedrine, a cardiac stimulant, which is also used in several over-the-counter asthma and allergy medications to open the airways. Interestingly, drugs with similar clinical properties are used today for supportive care of hospital patients infected with hantavirus.

A second plant depicted in the sandpainting, according to Muneta, is awe'e'tsa'a'l. The historical literature suggests this evergreen plant was used by some Native Americans in a cold medication and, when mixed with green branches, sagebrush, and juniper, could loosen the patient's mucus.

Among the non-Navajo people attending the meeting of healers in 1993 was Ron Voorhees, a physician and deputy state epidemiologist with the New Mexico Department of Health. "It became clear from that gathering that the Navajo knew just about everything about the virus and all we really added was a name," Voorhees told me. "Before the CDC identified the virus, the elders were saying, 'We've had this before! After wet winters. When there were a lot of mice.' That's essentially what all the research on the virus would later show. We've got DNA sequencing, so we can do all sorts of things and trace the evolutionary history of the virus. But the Navajo had the long history of observational epidemiology, which is pretty much what we used until computers made more complicated statistical analysis possible. The Navajo looked at people who got the disease and compared them with people who didn't; then they drew conclusions about how the people got it.

"Epidemiology is little more than structured observation," Voorhees continued. "They did their own risk-factor analysis. Navajos have a highly evolved culture, in which careful observations over many generations add up to a substantial knowledge base. One reason, probably, why they reached the same conclusion with fewer tools is that they have a much broader view of interconnectedness than we do. They are far less dependent on rigid linear connections, and they see connections in daily life that we can see only through statistics. They especially understood the basic ecology of the disease, something the CDC and the rest of us had no notion of until we heard the elders speak at the meeting."

5.

Robert Parmenter, a professor of ecology at the University of New Mexico and director of the university's long-term ecology research program at Sevilleta Research Field Station, doesn't put much stock in Navajo claims to know so much about the disease, but he puts a lot of stock in modern science and its conclusions about the origins of hantavirus. As leader of a study of deer mouse populations in the Four Corners area that has been going on for more than a decade, he has a lot of science to take stock in.

"The mouse data we had been collecting at our facility south of Albuquerque turned out to be invaluable because it showed fluctuating mouse populations over a very long time," Parmenter explained during my visit to the campus. "The spring of 1993 saw a huge explosion in populations. In an average year, perhaps one or two of every ten box traps we set out would catch one of the rodents. But in the spring of 1993, 90 percent of the traps were full by morning. We also kept very precise weather data, so it was easy to demonstrate, vis-à-vis the occurrence of hantavirus, that mouse populations always increased after unusually high winter and spring precipitation."

A second site, on the Navajo reservation nearly 200 miles north of Sevilleta, had seen unusually heavy rains, and surveys there showed an increase in mice as well as in cases of HPS. A third site, in Moab, Utah, turned out to offer a scientifically convincing point of comparison because it had not rained there during the year before the outbreaks elsewhere. In Moab, the density of mice remained comparatively low, and no cases of HPS had been reported.

"You can draw three conclusions from this data," Parmenter said. First, the rains in 1992 and early 1993 caused a dramatic increase in mice. During the appearance of El Niño in the summer of 1991, deer mouse population densities in New Mexico increased from about 15 mice for every ten acres to more than 75 per ten acres eight months later. By the spring of 1993, there were about 100 mice per every ten acres. Second, the initial human cases of HPS directly followed these increased densities of mice. "If you put these two together," Parmenter went on, "you come to the conclusion that increased winter and summer rain is associated with outbreaks of hantavirus."

"What caused all the unusually heavy rains?" I asked.

"El Niño. When the rains came, so did the sickness. When the rains left, the sickness left too."

But the picture turned out to be more complicated. In 2000, researchers from Johns Hopkins University completed a more precise analysis of precipitation data during the El Niño years. They discovered that even though rainfall was above normal in many areas, it was normal around the homes where the victims became infected. The rain nevertheless played a critical role, Parmenter explained. The deer mouse populations had exploded in the areas with unusually heavy rain, and then the mice spilled out of the canyons and traveled into secondary habitats, such as around houses, trailers, outhouses, and other places—the very places where people became infected.

6.

By the autumn of 1993, the snakeweed had become mounds of saffron flowers across the Plateau. The latest HPS outbreak there seemed to have vanished as quickly as it had arisen. The slender stalks of the downy grasses bent in the autumn breeze as their seed heads faded from green to purple and then brown and the plants approached their winter death, illustrating, in a sad and incongruous way, what the Navajo have always said: "In beauty it is done; in harmony it is written. In beauty and harmony it shall so be finished."

Florena's and Merrill's lives were over, but the disease would come again in ensuing years, claiming other members of their community and those far beyond. In fact, after the 1993 outbreak, the trail of discovery was just beginning. An increase in El Niño activity in 1999 again increased rainfall in the Four Corners area—followed by an upsurge in HPS. Researchers had also learned that the disease was not only a Four Corners phenomenon. By the summer of 2002, a total of 318 cases of the newly recognized hantavirus pulmonary syndrome had been identified in thirty-one states, including several in the Four Corners area. More than a third of the victims died. Although the vast majority of victims of the newly described disease turned out to be people who lived elsewhere in the United States, the disease was yet another instance of what to the Navajo has long been undeniable—human health and the fate of the natural world are inseparable.

$\cdots 6 \cdots$

A Virus from the Nile

1.

It probably happened in August. Beyond that, no one can say when the tiny brown wisp settled upon Enrico Gabrielli's body. The sixty-year-old cherished summer evenings among the red geraniums and purple cosmos in his garden, in the Italian neighborhood of Whitestone in Queens, New York— and never more so than in the summer of 1999. In July the temperature broke ninety-five degrees for eleven straight days—the hottest month ever recorded in the city.

On Wednesday, August 11, the gray-haired Gabrielli return-ed from his job at a mannequin factory in Elizabeth, New Jersey, and complained of fever and chills. His wife, Caterina, suspecting the flu, handed him two aspirin tablets and sent him to bed. He shivered and sweated throughout the night.

By the time Gabrielli was admitted to the intensive care unit of Flushing Hospital Medical Center the next day, he was feverish, disoriented, and unable to move. His strength rapidly faded. He began having trouble breathing and was put on a ventilator. A few days later, as he lay beneath dozens of get-well cards taped above the bed, Gabrielli opened his eyes and spoke. His 104-degree fever had broken. He had lost more than twenty pounds, which made his once-full face gaunt. Over the next few weeks he grew stronger, though he still could not walk on his own and relied on a catheter to urinate. Although he now walked with a cane, the life-threatening phase of his mysterious illness had passed, and Enrico Gabrielli, the first known victim of West Nile virus in the Western Hemisphere, had lived to tell about it.

On August 15, four days after Gabrielli began experiencing symptoms, an eighty-year-old man who lived a few blocks from the Gabriellis fell ill. Most evenings that summer, he and his eighty-two-year-old wife had sat outside their home, talking to each other and to passing neighbors. The annoying whine of jets passing overhead from nearby John F. Kennedy International Airport and LaGuardia Airport was a part of life. So was the sweet maritime scent of brine from the nearby marshes. Gray herons, egrets, gulls, and other shorebirds frequently passed over the neighborhood, making the skies above Queens a living diorama on the history of flight, ranging from the ultra-sophisticated structure of herons to crudely shaped modern aircraft. Aside from a manageable heart condition, the former World War II sergeant had been active and healthy. His wife believes that the mosquito bit him one August evening as he relaxed outside in his armchair.

Many houses in the neighborhood lacked air-conditioning. Evenings outside, always a favorite summer pastime in
northern Queens, were a necessity that year. April, May, and
June had been the driest stretch in more than a hundred
years. Newspapers carried headlines about the drought and
the heat, which killed more than a hundred people from the
Midwest to the East Coast.

"Is something new and different going on with the
weather?" asked science writer William K. Stevens in the
New York Times. The article said that the heat wave was part
of a fifty-year trend toward hotter summers in the region and
that heat waves and droughts could become more frequent.
Meteorologists attributed the drought to the naturally shifting warm-cold cycle of surface temperatures in the Pacific
Ocean linked to El Niño and La Niña years, while climate
scientists suspected that the increasing severity of shifts in
the recent past has been caused by global warming.

The drought made misery for humans, but it benefited
one of the most common biting insects in Queens. The
northern house mosquito, Culex pipiens, often thrives during droughts. After getting a blood meal, a female mosquito
deposits eggs in wastewater, which is laced with organic
nutrients. Because of the drought, the city sewers had not
been flushed by rain in months, creating the organically rich
standing water the egg-laden females preferred. When the
mosquito eggs hatched, the dry heat above ground tended to
confine the emerging insects to the humid sewers. On
August 5, the first rain in weeks fell on Queens, helping to

liberate the blood-seeking wisps from their subterranean lairs. At dusk they fanned across the borough.

Although they naturally prefer birds, the mosquitoes bite humans and other mammals as well. There are actually two beneficiaries of their blood meals: the mosquitoes themselves and any viruses they might harbor—West Nile virus in this case. The virus needs a living being—a host—within which to replicate. Each time the mosquito bites a bird, the virus within the mosquito has an opportunity to move into a new host. Without such living incubation chambers, or reservoirs—and a means of traveling to new ones—a virus would quickly die out because its life within any particular animal may be brief. By using a mosquito as a vector, the virus can quickly spread and become established in millions of birds through a process appropriately known as amplification. It is a diabolically effective system. Mosquitoes also pass the virus to people, where it can replicate in the brain. That was the case with Gabrielli and his elderly neighbor.

On Saturday, August 12, a week after the half-inch downpour, the elderly neighbor came in from mowing the front lawn and complained of extreme fatigue. "It was the first time he'd ever complained," his wife said. He wouldn't eat. He vomited and went to bed. The next morning, his wife expected him to be up as usual by four or five o'clock, banging around the kitchen and making coffee. Instead, he could barely open his eyes. When he slid his arm over her waist, she noticed that his hand was hot. "Somehow he couldn't move right," she remarked. She canceled plans to visit their daughter, who instead came to Queens that afternoon. When she arrived, she saw that her usually neatly dressed father had tucked in only half his shirt. He was laboring to speak in single syllables, and later in the day he collapsed in a chair. An ambulance drove him to Flushing Hospital, where doctors were able to revive him. He was admitted to the intensive care unit, only a few beds from where Enrico Gabrielli lay, but his liver and kidneys began to fail and he suffered a heart attack. Soon thereafter, he died. The former soldier was buried on Long Island—the second known victim and the first fatality of the mysterious disease.

By August 23, three more patients with neurological symptoms had been admitted to Flushing Hospital. Deborah S. Asnis, a staff physician and infectious disease specialist, telephoned the New York City Department of Health and Mental Hygiene to report the unusual cluster of illnesses. After discovering that nearby hospitals had admitted another five patients with similar symptoms, the health department contacted the Centers for Disease Control and Prevention. The next day, an official from the CDC's Epidemic Intelligence Service flew to Queens to interview surviving patients, comb medical records, and visit the homes of the afflicted in an attempt to identify the disease and determine how it spread.

By early September the CDC had come up with an answer. The mayor of New York City held a news conference in Queens to announce that a disease known as St. Louis encephalitis, caused by a mosquito-borne virus, was responsible for the human deaths. St. Louis encephalitis, named for the city where, in 1933, it was first identified, had never before been seen in New York City. This was a public health emergency, and Mayor Rudolph Giuliani promised to "do everything we can to wipe out the mosquito population."

New York City's Office of Emergency Management set up a command post at 138th Street and 11th Avenue, near the Gabriellis' house. The city mobilized eleven spray trucks, five helicopters, and an airplane to douse the city with pesticides. Police officers cruised neighborhoods, warning residents over loudspeakers to remain inside with their windows closed. An advertising campaign called "Mosquito-Proof New York City" was launched.

New York City had last seen a mosquito-borne disease during a yellow fever outbreak in the early 1800s. Most modern New Yorkers could not grasp the idea of a common mosquito injecting people with a potentially fatal virus. A resident of a neighborhood adjacent to Whitestone also worried about the mental health of her children, who had developed a paralyzing phobia of flying insects. "If they see a fly," she said, "they think that they are going to die."

2.

Months before the first human death, hundreds of crows had begun dying in Queens. Some had been found within blocks of Flushing Hospital. One woman found a disoriented crow hobbling in her garden. At Bayside Animal Clinic, veterinarian John Charos treated more than fifty ill crows. Half of them ultimately died. A security guard at Fort

Totten, a 163-acre government property in Queens, found dead crows all over the base and likened it to "a plague." Crows and other birds were also dying in the Bronx, across the bay from Queens. Near 198th Street and Briggs Avenue, a passerby happened upon four dead pigeons. Forty dead crows were found near the Bronx Zoo, where a captive cormorant, three Chilean flamingos, a pheasant, and a bald eagle also died.

Many people blamed the rash of bird deaths on the drought. Ward Stone, pathologist at the New York State Department of Health in Albany, said it was the worst die-off of crows in thirty years. According to the drought theory, the heat had driven earthworms, insects, and other sources of food deeper into the ground. As the crows dug, they encountered persistent toxins, such as DDT, that had contaminated the soil a half-century earlier, when the pesticide was commonly used.

Tracey McNamara, a veterinarian and head pathologist at the Bronx Zoo, questioned the drought theory. Crows were hardy, adaptable, and resourceful birds, she thought; why would a drought affect them more than other birds? Furthermore, a drought would not have directly affected captive zoo birds, which had all the food and water they needed. The CDC's diagnosis of St. Louis encephalitis didn't explain the bird deaths, either, she realized, because birds generally aren't susceptible to St. Louis encephalitis.

McNamara's hunch was that a different—perhaps new—virus was responsible, even though viruses that killed both birds and people were few and far between in North America. One candidate, at least in theory, would be eastern equine encephalitis, or Triple E. This disease, which also attacks the brain, can kill not only birds and people but also horses and individuals of other species. The Triple E virus was known to be especially lethal to emus, ostrich-like birds from Australia. Yet the Bronx zoo had a number of emus, and they remained healthy even as the other birds died during the mysterious outbreak. This fact alone all but ruled out Triple E as the culprit.

Another observation that weighed in favor of a new virus, McNamara believed, was that all the birds stricken by the disease at the Bronx Zoo were native to the Western Hemisphere. Did this mean that the virus had moved here from another part of the world and was killing only birds whose immune systems were unprepared for this exotic invader?

On September 9, two more flamingos died at the zoo. While taking blood from one of them, McNamara's colleague accidentally stuck herself with a contaminated needle. If the bird and human deaths were related, McNamara realized, the technician's life could be in danger. That day she called the CDC to ask about her colleague's exposure and to suggest that the human and bird die-offs were related. In doing so, she was discounting St. Louis encephalitis as the cause and thus challenging the CDC's diagnosis. The latter is not something a veterinarian—or anyone else, for that matter—usually does.

3.

Every autumn, clouds of white storks move over Israel as they migrate from breeding grounds in Europe to wintering grounds in Africa. A more direct flight to Africa would carry them across the Mediterranean Sea, but these heavy birds rely on thermals—currents of air that rise from warming land—to keep them aloft and carry them on their journey, forcing them to avoid large bodies of water, even at the expense of a longer migration route. The summer of 1998— a year before the virus struck in the United States—was the hottest in thirty-five years in Israel. Temperatures along the migration route regularly reached 100 degrees Fahrenheit and occasionally soared to 116 degrees. Winds gusted to thirty miles per hour. Unable to navigate the winds or endure the extreme heat, tens of thousands of the stressed birds that had hatched in Europe landed in Israel.

One flock of 1,200 birds set down at Eilat, in Israel's southern tip, near the Red Sea. Farmers in the region soon began finding dead storks in their fields. Not long afterward, hundreds of domestic geese in villages around the country mysteriously succumbed to an unknown disease. Many of them had neurological abnormalities: they could not stand or keep their balance. Israel's government tested a number of wild storks and the geese, and the brain of a dead goose yielded West Nile virus. Common throughout Africa and, more recently, in Europe, the virus had visited Israel in the 1950s and the late 1970s—but not again until the summer of 1998. The young European storks may have reintroduced the virus into Israel, where it then infected domesticated geese. Perhaps under normal conditions, even infected storks would have remained healthy—some birds carry the virus without ill effect—but under the stress of a difficult migration, the storks fell ill. But not, apparently, before spreading the virus to mosquitoes in the area where they landed. The mosquitoes then could have easily infected the goose farms. This drama in Israel was unfolding 7,000 miles away from Queens, and a year before either a person or a bird there would fall ill from the disease.

4.

On September 9, 1999, when McNamara telephoned the CDC's Division of Vector-Borne Infectious Diseases in Fort Collins, Colorado, to express concern about her colleague's accident, her call was transferred to the chief of the Epidemiology and Ecology Section. McNamara asked whether the CDC would test blood samples she had taken from her colleague and the dead birds to see whether the same virus could be isolated from both. It was inappropriate, the official explained, for an institution concerned with human health to test birds' blood; indeed, since the viruses in birds and in people were different, the CDC thought it superfluous even to test the human blood sample. At a loss for what else to do, McNamara sent both samples to the National Veterinary Services Laboratories (NVSL) in Ames, Iowa. Several days later, an official at the NVSL called McNamara to say that an unusual virus had been isolated from both samples. The lab could not determine exactly what the virus was, beyond the fact that it was a member of the dangerous *Flavivirus* genus. Yet flaviviruses had never been associated with bird fatalities in the United States. Had a new one arrived or an old one mutated?

Definitive identification of the dangerous virus would require a secure laboratory to prevent human infection. Only a handful of such facilities existed in the United States. One of these was housed at the U.S. Army Medical Research Institute of Infectious Diseases (USAMRIID), at Fort Detrick, Maryland, the military's main biological warfare laboratory. McNamara happened to have a friend who worked there, and that person agreed to test the samples. A researcher at the New York State Department of Health's laboratory in Albany also agreed to run further tests on the samples forwarded by the NVSL.

Meanwhile, McNamara telephoned John T. Roehrig, chief of the CDC's Arbovirus Diseases Branch, and told him that the NVSL had isolated something that looked very much like a flavivirus. She also pressed her concern that the bird and human deaths were linked—implying that the virus was indeed something new in this part of the world. Confirmation that a flavivirus was killing birds in the United States would be a historic and ominous finding. The CDC's response was still the same: the agency insisted that the human and avian deaths were not related and thus there was no logical reason for them to test bird samples. McNamara would have to await the findings from her friend at USAMRIID.

On September 23, the telephone rang in McNamara's
Bronx Zoo laboratory. Several senior scientists from the CDC
were on the line. Roehrig asked McNamara to ship frozen
samples directly to the CDC that night. He said there had
been some confusion with the samples the NVSL had sent
them earlier. The callers said little else. To McNamara, this
spoke volumes.

"Is it okay to be working with the virus here?" she asked in
alarm.

Because she wore a mask and gloves and worked under a
special ventilated hood, it was probably okay, she was told.

When the conference call ended, McNamara quickly
telephoned several friends in the know and learned that the
Fort Detrick lab had definitively ruled out three possible
candidates for the outbreak: eastern equine encephalitis,
western equine encephalitis, and Venezuelan equine encephalitis. The tests further suggested that St. Louis encephalitis,
the CDC's original diagnosis, wasn't the cause either.
USAMRIID had reported its findings immediately to the
CDC and had continued searching for the identity of the
mystery virus. Meanwhile, officials at the CDC, finally
beginning to grow alarmed, agreed to test the samples
McNamara shipped them.

On September 30, 1999, the CDC issued a press release
announcing it had now "made the link between the West
Nile–like virus found in birds in New York City and the
ongoing human encephalitis outbreak in the area," thus confirming McNamara's hypothesis. Geneticists immediately

began comparing the New York strain with strains from Africa, Europe, and elsewhere to determine where the New York strain had originated. They soon found that it matched a sample isolated from the brain of the dead goose in Israel, a strain common throughout the Middle East.

West Nile virus was first discovered in the 1930s, when it was isolated from a woman living on the west side of the Nile River in Uganda. Since that time, birds migrating from Africa had spread the virus along their migration routes throughout much of the Middle East and Europe. But there were no bird migration routes from those countries to the East Coast of the United States. How, then, had the virus traveled thousands of miles from the Middle East to the borough of Queens?

5.

The thousands of Queens residents sitting on their porch stoops and patios the summer of 1999 had no particular reason to realize they lived near one of the greatest crossroads—for both people and birds—the world had ever known. Each month, some 11,000 overseas flights to Kennedy International Airport bring more than 2 million people through Queens. More than 20 million overseas passengers disembark there annually. That does not include the almost 4,000 horses and thousands of exotic birds, turtles, and fish and other animals that legally pass through JFK every year. Hundreds more animals—perhaps thousands—evade the quarantines and inspections set up to keep out imported diseases. And no one attempts to account for the numerous small six-legged, winged, or tiny crawling stowaways from the aircraft's cabins and pressurized holds and from the bodies of the passengers. If Queens is a cultural melting pot, it is also one gigantic petri dish. In its own way, Queens rivals some of the world's other great interspecies crossroads, such as Guangdong Province of southern China, the epicenter of the SARS outbreak.

Kennedy International Airport also lies along the Atlantic Flyway, a major migration route for birds flying between the Americas. In fact, runway 22L juts into the 10,000-acre Jamaica Bay Wildlife Refuge, which is visited every year by millions of birds from Mexico, South America, the Caribbean region, and the far north. Many people forget about the vast network of rivers, wetlands, and shorelines that surrounds New York City. The birds have not.

In spring and early summer, the songs of warblers, vireos, swamp sparrows, goldfinches, and eastern bluebirds arise from the trees and forest patches, and snowy egrets, black-crowned night herons, sandpipers, belted kingfishers, and great blue herons fill the wetlands. The occasional great cormorant, green-winged teal, and red-breasted merganser wanders through. In the winter of 1998, a rare sighting of a European widgeon, perhaps from a far-flung Iceland flock, was made in Queens. Escaped parrots and other tropical birds are also occasionally documented in the borough.

Birds and people are not the only species passing through the region. Monarch butterflies migrate through New York

City in autumn, feeding on life-giving milkweed in wayward urban lots and along roadsides. The monarchs are destined for New Jersey's Cape May, where they congregate by the thousands before continuing their patient journeys, on breezes or one wing-stroke at a time, to Mexico.

6.

In late August 1999, when Tropical Storm Floyd dumped nearly five inches of rain on New York City, the historic drought of that year became a memory. By mid-September, when the year's final case of West Nile fever was diagnosed, seven of the fifty-nine people hospitalized with the virus in New York City had died. Enrico Gabrielli, home from rehabilitation for several weeks, walked with a cane.

But a large part of the West Nile virus mystery remained. How had it arrived in the New York region in the first place? Perhaps a person bitten by an infected mosquito in the Middle East had carried the virus to Queens, only to be bitten by another mosquito in New York. Perhaps that mosquito then fled across the parking lot outside the international arrivals terminal and disappeared into the refuge at Jamaica Bay, where it spread the virus to other birds and mosquitoes, many of which could have ended up in the neighborhoods of Whitestone and Flushing. Perhaps an infected mosquito arrived in an aircraft cabin or cargo hold. Or maybe one of the numerous parrots, parakeets, or lovebirds smuggled through New York each year was infected. It is conceivable that, by some strange anomaly of normal migration, a bird infected with the virus in Europe or in Africa passed the infection to a bird that migrated to Queens. Given the right conditions, a single infected bird could pass the virus to a mosquito. The insect, in turn, could quickly infect other birds, igniting a rapid outbreak of the virus in both birds and people.

As autumn arrived, many birds left the New York region for their wintering grounds. A world away, a new skyful of white storks rode thermals above the hot sands of the Middle East and Israel toward Africa, as they have done since before the time of Abraham. The last wave of monarchs, propelled by the laggard hurricane winds, departed New York City and environs for forests in Mexico three thousand miles away.

Migrating monarchs, white storks en route to Africa, cooling waters off Peru, winds across Arabia, an empty lawn chair in Queens, and a fresh grave on Long Island. A black-crowned night heron lifted from the waters off Whitestone, circled as if on a designated flight path, and disappeared into the night.

7.

Many of the birds that left the New York City region in the autumn of 1999 migrated south along the Atlantic Flyway, some fanning into wetlands in South Carolina, Georgia, and Florida. Although the virus was especially lethal to crows and jays, more than a hundred other species carry the virus. Many of these may have survived an infection, enabling them to carry it far and wide.

In July 2001, the first confirmed cases of West Nile fever occurred outside the New York–New Jersey metropolitan region when seventy-three-year-old Seymore Carruthers of Madison County, Florida, fell ill. A short time later, a sixty-four-year-old woman, also from Madison County, came down with the disease. "The virus is spreading," commented Steve Wiersma, chief of Florida's Bureau of Epidemiology. "We might slow it but we can't stop it. Nothing can stop it. The ecology is here. The birds are here, and the people are here. Of course, the mosquitoes are here and will always be here."

The viral plume soon stretched all the way south to Marathon, in the Florida Keys, when a vacationing seventy-three-year-old woman from Sarasota, suffering from confusion, swollen lymph glands, headache, and a high fever, was diagnosed with West Nile fever. She recovered and was released from the hospital several days after being admitted. By the end of 2001, the virus had infected people in ten eastern states, from Massachusetts to Florida, over an area of half a million square miles, and it had been detected in birds across the eastern half of the United States.

No one expected the virus to stop there, not least of all David Rogers, a professor of ecology at Oxford University who had been tracking the virus since its arrival in the New York City area. Working with colleagues from the National Aeronautics and Space Administration (NASA), Rogers developed risk maps to predict where the virus was likely to strike next—thus potentially alerting people in the disease path to take precautions. After the Florida outbreak, Rogers began feeding into a computer at Oxford satellite images of ground vegetation, temperature, and other information suggestive of good mosquito habitat. On this data he then superimposed the coordinates of areas in which infected birds had been found. By early 2002, the NASA team had identified Louisiana as a potential trouble spot. True to its prediction, by late summer the epidemic had struck there. Fifty-eight people fell ill. West Nile had also arrived in Mississippi and several nearby states.

The virus struck Louisiana with such fierceness that some speculated it might have mutated into something more virulent than the strain from New York. For one thing, the Louisiana outbreak seemed to be striking a higher percentage of young people than had West Nile outbreaks in the previous three years. During the virus' first two years in the United States, the average age of patients was about sixty-six years; in 2001 it was even higher, seventy. But during the initial 2002 outbreak in the Gulf of Mexico region, the victims' average age was in the upper fifties. Of the fifty-eight cases, twelve of the victims were between forty-five and fifty-nine years of age, and nineteen were younger still. Was the shift in age a coincidence, or had the virus undergone an ominous mutation that gave it the power to overwhelm the relatively healthier immune systems of the young? Time would tell.

"The peculiarity about West Nile virus," David Rogers told me, "is that it appears to be supported by at least thirty species of mosquito vectors in the United States and at least

eighty species of bird hosts, not to mention some other animals. Normally diseases—even viral ones—have fewer hosts
and vectors. Even the relatively close cousins of West Nile
virus, such as yellow fever, have far fewer. The greater the
number of vectors and hosts, the more likely a disease is to
spread within a new continent. In three years West Nile
virus in the U.S. has gone from zero to thirty-four states.
That's a record by any standard."

Rogers pointed out that in 2002 alone there were 4,161
documented cases, with 284 deaths, in forty-four states and
the District of Columbia. The Midwest was particularly hard
hit, with nearly 2,000 documented cases in just Illinois,
Michigan, and Ohio. "This must make West Nile fever one
of the most important vector-borne diseases in the entire
United States—if not *the* most important—all within three
years of its first appearance in New York," he said.

"Because it's carried by so many insect and bird species,
it's extremely difficult to predict just where it will show up
next," Rogers continued. "Our maps can tell you where not
to take your granny on vacation or to have a picnic. My prediction is that, given the mosquitoes there, West Nile virus
will appear in California and become established in the San
Joaquin Valley in the not too distant future." Less than two
months after Rogers said this, the first human case of West
Nile virus in California was identified, although the victim
had apparently been infected by a plane-borne mosquito
from the East Coast. "If there's not a mosquito already in
California capable of transmitting the disease," Rogers

warned, "the Asian tiger mosquito, which is a vector, will likely become established there soon."

Outside the United States, meanwhile, West Nile virus also appears to be on the move into new areas. "There seem to be pockets of West Nile–like viruses in southern Europe that are not necessarily infecting people, but we are finding them in horses and in wild birds," noted Mertyn Malkinson of Israel's Kimron Veterinary Institute. She also said there are signs that the viruses are generally moving northward in Europe.

Assisted by migrating birds, international travel, and perhaps a warming climate, a so-called African virus is becoming global. Like a smoke plume swept into a wind, West Nile virus is spreading.

We humans are prone to forget that our peregrinations are forever charted on the larger map of the biological world. Wherever we and other animals travel, the microscopic life upon and within us also goes. Sometimes we leave microbes behind, and sometimes we bring new ones back with us. This realization is hardly a reason to remain at home. But it is cause for concern and a call for increased awareness of how quickly new diseases can emerge and travel in the age of globalization.

"One expects viruses to travel," Rogers concluded. "They always have, especially when migrating animals are part of the equation. But the rapidity of the spread of West Nile virus is unprecedented. One cannot say exactly where it is going to stop."

SARS and Beyond

It is only a matter of time, many epidemiologists warn, until another epidemic on the scale of the Spanish influenza outbreak of 1918–1919, or the current HIV/AIDS pandemic, sweeps across the globe. The National Academy of Sciences' Institute of Medicine recently cautioned:

> Today's outlook with regard to microbial threats to health is bleak on a number of fronts. . . . Pathogens—old and new—have ingenious ways of adapting to and breaching our armamentarium of defenses. We must also understand that factors in society, the environment, and our global interconnectedness actually increase the likelihood of the ongoing emergence and spread of infectious diseases.

So it shouldn't have been a complete surprise when, in late 2002, a previously unknown virus struck people in the form of severe acute respiratory syndrome, or SARS. David

147

L. Heymann, executive director of communicable diseases at the World Health Organization in Geneva, Switzerland, declared SARS "the first severe and easily transmissible new disease to emerge in the 21st century." It has now joined a growing list of disturbing new epidemics that already included HIV/AIDS, Ebola hemorrhagic fever, Lyme disease, and a host of others—some mild, others fatal.

This recent outbreak began, according to the Centers for Disease Control and Prevention, in Guangdong Province in southern China in November 2002, when dozens of people there began to experience headaches, muscle soreness, and dry coughs that quickly deteriorated into life-threatening pneumonia. Within months, the illness had spread throughout Guangdong, where government authorities, fearing social unrest and the loss of tourism, tried to keep the outbreak secret. No medical statistics were released, and journalists were prohibited from reporting on the deadly epidemic.

On February 21, 2003, Liu Jianlun, a sixty-four-year-old kidney specialist from Zhongshan Hospital in Guangdong, traveled to Hong Kong, where he stayed in room 911 at the Metropole Hotel. He had a fever and had not felt well for five days, but when he began to have trouble breathing, he went to Kwong Wah Hospital in Hong Kong. Suspecting he had contracted the highly infectious illness, he asked to be put in an isolation unit. He died several days later. China's secret would be no more.

Others who had stayed at the Metropole during the doctor's time there began to fall ill as well, including Kwan Sui-chu, a seventy-eight-year-old woman from near Toronto. She and her husband returned to Canada, and their forty-three-year-old son became infected. Kwan Sui-chu soon died, as did her son eight days later. Hundreds of people in the Toronto area—mostly health care workers and patients at the hospital—soon came down with SARS, with dozens of deaths. Other guests from the Metropole unwittingly spread the virus to Singapore and elsewhere in Asia.

Toward the end of February, several health care workers and patients at Vietnam's Hanoi French Hospital suddenly became seriously ill, and a doctor and nurse there quickly succumbed to pneumonia. Suspecting that the mysterious Hanoi outbreak might be related to the earlier pneumonia epidemic in Guangdong, the World Health Organization sent physician Carlo Urbani, director of infectious diseases for WHO's Western Pacific Region, to investigate. While there, he examined Johnny Chen, an American garment merchandise manager based in Shanghai who had been treated by the doctor and nurse who died. Chen, it turned out, had also been a guest at the Metropole during the time of Liu Jianlun's stay. On February 24 he had flown on to Hanoi before falling ill. Taking note of the illness' virulence, contagiousness, and particular effect on the hospitalized victims' lungs, Urbani recognized the disease as something new and terrible. Chen soon died, and Urbani himself died of the pneumonia a few days later.

By March 2003, the Chinese authorities had begun to reveal the extent of the outbreak in Guangdong. But now the disease was already sweeping through Hong Kong, infecting hundreds and killing dozens. It had begun to invade Beijing and other cities. It had also arrived in the United States. The disease spread readily from person to person through coughing, sneezing, and other means. It was also quickly disseminated around the globe by jet aircraft. One traveler flew from Hong Kong to Frankfurt and Munich, then on to London, back to Munich and Frankfurt, and then again to Hong Kong, apparently before even suspecting he had contracted a new disease.

On March 12, WHO declared the new virus "a worldwide health threat." Strict isolation of suspected cases and extreme precautions by health care workers eventually began to slow the spread in many countries, but the virus would remain out of control for weeks in China, where dozens of new cases were being reported every day. At the time, epidemiologists suspected that about 7 percent of patients died. At that rate, if one in ten people in China were to become infected, more than 7 million would die. If one of ten people worldwide came down with the illness, the death toll could exceed 30 million—a toll comparable to that of the Spanish influenza epidemic. Researchers soon learned that the fatality rate, at least in Hong Kong, was closer to 15 percent for those under sixty years of age and more than three times that for those older than sixty.

Genetic analysis showed that the SARS virus had come from a non-human animal—not a complete surprise, given that coronaviruses commonly infect other animals. In cattle, for example, a coronavirus has been associated with respiratory disease during the stresses of transport—hence its name, "shipping fever"—and a cattle type of coronavirus has even been associated with illness in people. The virus that causes common colds in people is also a form of coronavirus. Nor was the apparent origin surprising given that 75 percent of new human diseases identified in the past thirty years are closely related to wildlife or domestic animals.

In May, researchers at the University of Hong Kong, together with the Shenzhen Centre for Disease Control and Prevention, found a virus closely related to the SARS virus in masked palm civet cats, a weasel-faced tree-dweller native to Asia and consumed as a medicinal in China in the belief that it helps people withstand cold weather. The research suggested that the virus possibly jumped from that species to humans. Scientists also found evidence of a SARS-like virus in two other mammal species.

The World Health Organization, while cautioning that this evidence does not prove that the virus had jumped from these animals, nevertheless cautioned that "persons who might come into contact with these species or their products, including body fluids and excretions, should be aware of the possible health risks, particularly during close contact such as handling and slaughtering and possibly food processing and consumption."

China later temporarily banned trade in the cats, and the United States prohibited their import. Fortunately, prompt and strict isolation of patients eventually curtailed the spread of the virus. But still, by the end of July 2003, more than 8,000 cases had been identified in twenty-nine countries. Nearly 10 percent of those who contracted the virus—about 775 people—had died.

After a lull, in late 2003, China's Ministry of Health reported four new cases, all in Guangdong. All the patients recovered, however, without developing severe symptoms. The first months of 2004 saw a handful of mild cases in China. Had the virulence of the disease waned—and if so, why?—or was the old virus merely lurking? Many details about the exact origin of the virus remain unknown, as does its future as a human pathogen.

How might such a jump from civet cats and other species have occurred?

A coronavirus from one species would almost certainly have had to undergo a series of genetic changes before it could successfully infect a new species. Such changes can occur in two basic ways: the virus either incorporates genetic material from another virus or undergoes a spontaneous change in its own genetic makeup.

In the first scenario, the virus—or bacterium, for that matter—may incorporate genetic material from a different microbe, thereby creating a new type. This is one of the means underlying the regular emergence of new strains of influenza from Guangdong and its environs. The region has given rise to the 1957 Asian flu pandemic, the 1968 Hong Kong flu pandemic, and, more recently, "bird flu," a potentially dangerous disease that was initially halted by immediate reporting and extermination of millions of potentially infected poultry before they reached Hong Kong markets. But few epidemiologists believed that bird flu, also known as H5N1, was actually gone for good. Once established, a virus is nearly impossible to eliminate from an animal reservoir. In the case of influenza, this reservoir included wild aquatic birds (ducks especially) as well as domesticated ducks, chickens, and pigs.

After the 1997 Hong Kong flu outbreak, bird flu broke out sporadically in China and the Netherlands. By late 2003, H5N1 had returned in full force, in a revival that, in its extent and virulence, the World Health Organization called "historically unprecedented."

The new bird flu outbreak was first reported in December 2003 in poultry in Korea, and it quickly spread to poultry in several other countries, including Japan. In the second week of January 2004, the virus killed almost 6,000 chickens on a single farm in the town of Ato in Yamaguchi Prefecture. Eggs from the farm were recalled, and ultimately more than 28,000 birds were exterminated and buried in an attempt to prevent further spread. Hundreds of thousands of domesticated fowl were subsequently exterminated in Vietnam, China, Cambodia, Indonesia, and Laos. The toll on domesticated fowl in other countries was grim, forcing many large producers and small farmers alike into financial ruin. Different strains of bird flu emerged at the same time in Canada, the United States, and elsewhere. Fortunately, though sometimes fatal for poultry, these types were incapable of infecting humans with serious disease.

The toll on poultry farmers was terrible enough, but what worried officials at the World Health Organization even more was the potential for the virus to spread easily to people. By the end of February 2004, for example, ten people in Thailand had been infected, with seven deaths. Vietnam had twenty-three human cases, with fifteen deaths. In all these cases, the human victims apparently had been infected directly by poultry. If the bird flu virus ever readily adapted to humans, it could begin to spread from person to person.

The same farms where various strains arose could offer bird flu every opportunity to evolve into a true human killer. Pigs, for example, can be infected with both human and bird influenzas simultaneously. If cells in a pig's trachea were infected with both viruses, the avian form of the virus could pick up genes from the human form. A hybrid strain might then arise with the deadliness of bird flu and the contagiousness of human flu. Influenzas are talented gene swappers, and the potential for a pandemic strain arising is real.

Why do many new diseases seem to emerge from southern China? Because the various viral influenza strains could intermingle as they passed from wild birds through various farm animals, the sprawling agricultural enterprises throughout China have become a global influenza factory. It was for good reason that southern China has become known as the epicenter of flu viruses. The Asian flu epidemic of 1957, the Hong Kong flu epidemic of 1968, and now the H5N1 epidemic among poultry had all been born there.

Another reason is the close intermingling there of people with many other species, which gives viruses of one species numerous opportunities to undergo this kind of genetic exchange. For example, pigs are sometimes raised in the same living quarters as their owners, and in some cases chickens are kept directly above pigsties. The microbes in the chicken manure end up in the intestines of pigs, where they may further evolve in this new environment. Runoff from pigsties drains into ponds where shrimp and grass carp are raised. Ducks and other birds, which also harbor viruses and bacteria, frequent these ponds and contribute their feces to the variety. Human excrement often ends up with that of the other animals.

The problem of accelerated genetic swapping is not limited to domestic animals. Parts of southern China have huge markets that supply numerous species of wild animals to the culinary trade—another rich venue where microbes can mix, match, and mutate. Chau Tau Market, which sprawls over several city blocks on the outskirts of Guangzhou, the province's capital, features an array of wild mammals (as well as cats and dogs), amphibians, birds, fish, and reptiles to supply the exotic culinary tastes of Cantonese chefs. All these species intermingle with their human handlers and butchers and those who later prepare the food from raw meat. The close contact between people and numerous other species makes the region a microbial melting pot.

Of course, China isn't the only region of the world where microbes are undergoing intensive genetic exchange because of the intermingling of people with nonhuman animals. But the vast interspecies crossroads of domestic animals and wild in Guangdong, along with its 75 million people, does make it an incubation chamber for new diseases. And the proximity of Hong Kong offers new infectious microbes a plane ticket to the rest of the world.

Although the coronavirus behind SARS is in a completely different family from the influenza viruses, it, like many other viruses, can evolve by similar means. And this brings us to a second scenario for genetic change. If a virus requires genetic changes in order to jump to humans, it may evolve not only by incorporating genetic material from another virus but also simply by undergoing a spontaneous genetic change that happens to allow it to infect people. Such mutations may have given the SARS virus the keys to enter the human species. The vast majority of mutations are inconsequential for human health. Given billions upon billions of such changes, however, a few are likely to ultimately benefit the virus—perhaps allowing it to infect a new species.

But even in the case of a natural mutation, people's actions may ultimately determine the fate of a new virus, just as they can facilitate genetic exchange. For example, dense human populations in constant contact with an array of naturally mutating viruses may increase the chance for a potentially deadly genetic creation to find people to infect. What's more, intensive agriculture and other conditions that create extremely high densities of various viruses—in waste ponds, for example—mean, as a matter of simple probability, that more random mutations will occur there. The increased numbers of mutations, along with almost unlimited possibilities for these microbes to find people, either directly or through other animals, increase the chance that another epidemic will be born.

As our attention is—understandably—diverted by frightening new illnesses such as SARS, old threats, such as antibiotic-resistant disease, which kills some 20,000 people every year in the United States, continue to grow. Some of the gravest threats remain the subtle genetic changes we foster in old scourges. Our attention may be focused on the newest epidemic, but death is often in the details of the old ones.

When a frightening new disease does come along, only a deep sense of bias and denial permits us to point reflexively at a phenomenon like bioterrorism as the cause while ignoring our own collective promulgation of the global environmental disruption that has given rise to so many new infectious diseases. We frequently spell out the dangers of deliberate genetic engineering even as we maintain what amount to giant genetic engineering laboratories—in the form of the intensive agricultural systems that create agents such as the one responsible for mad cow disease, which terrorize in their own way.

But even in an age of mounting epidemics there is hope. With early detection, many epidemics will be contained. For those that are not, medical technology may offer some respite. Changes in personal behavior alone can greatly reduce the risks of many diseases, such as HIV/AIDS. Mad cow disease was quelled simply by banning the feeding of meat and bone meal to cattle, allowing them to be herbivores again. As the overuse of antibiotics has caused many bacteria to become resistant, using antibiotics more sparingly may help roll back the tide of some untreatable bacterial infections.

Understandably, when faced with a deadly new infectious

disease, we typically don't rely on short-term solutions. We want first and foremost to protect ourselves and find a cure. Once that is done, it's easy to think that the problem has been resolved. And to guard against a next time, we want better vaccines, medicines, and other technologies. We want more effective government policies today.

But what about tomorrow and the new diseases that will inevitably arise?

Since pathogenic microbes will always be a part of the natural environment, we must find a way to live with our infectious adversaries rather than try to eliminate them. As biologist Paul W. Ewald argues, we must take actions that will cause infectious diseases to evolve away from virulence and toward more benign forms. It is a lofty but plausible ambition.

Although it is too early to say for certain, we may have done just that with SARS, according to Ewald. By quickly and effectively quarantining infected individuals, medical professionals prevented the virus from passing easily from person to person. If the virus "hoped" to survive, it had to evolve into a milder form—to keep its victims up and walking, so to speak—and continue to spread itself.

Did deliberate human actions force the virus to become milder? If so, wouldn't this solution be more effective over the long term than all the vaccines or drugs that might have been developed to combat the disease? If so, we overcame the initial epidemic not by eliminating the virus but by modifying it. This is not to say that SARS is forever tamed. But it suggests that approaches to fighting new diseases must address the human influences on their emergence and behavior.

But what of the deep ecological, demographic, and industrial roots of the surge in new epidemics? Will humans be able to ameliorate the upset predator-prey balances that have helped precipitate Lyme disease? With the urgent need for protein in some parts of Africa, can people reduce the massive consumption of bushmeat that may be predisposing humans to new forms of HIV? Will we be able to curb the effects of climate change that are causing some disease-causing organisms to spread to new areas or otherwise proliferate? Although we are learning much about the ecological origins of new diseases, it remains to be seen whether we will address these mounting epidemics at their roots. This will require more than devoting ourselves to new treatments and cures alone. It will also require curing the cause; and that means protecting and restoring the ecological wholeness upon which our health often depends.

Notes

Introduction

p. 1 *strange new disease in the city:* J. Steinhauer, "As Fears Rise about Virus, the Answers Are Elusive," *New York Times*, September 29, 1999.

p. 1 *in some cases were comatose. Several soon died:* D. S. Asnis et al., "The West Nile Virus Outbreak of 1999 in New York: The Flushing Hospital Experience," *Clinical Infectious Diseases* 30, no. 3 (2000): 413–418.

p. 1 *Black Death, which wiped out as much as one-third of Europe's population in the 1300s, or . . . Spanish influenza:* W. H. McNeill, *Plagues and Peoples* (Garden City, N.Y.: Anchor Press/Doubleday, 1976).

p. 2 *severe acute respiratory syndrome (SARS):* World Health Organization, "Severe Acute Respiratory Syndrome," *Weekly Epidemiological Record* 78, no. 12 (2003): 81–88; World Health Organization, "WHO Issues a Global Alert about Cases of Atypical Pneumonia: Cases of Severe Respiratory Illness May Spread to Hospital Staff," press release, March 12, 2003; L. K. Altman and K. Bradsher, "Official Warns of Spread of Respiratory Disease," *New York Times*, March 30, 2003; World Health Organization, Communicable Disease Surveillance

· · · · · · · ·

and Response, "One Month into the Global SARS Outbreak: Status of the Outbreak and Lessons for the Immediate Future," Update 27, April 11, 2003.

p. 2 *"We may be in the very early stages of what could be a much larger problem as we go forward in time"*: Centers for Disease Control and Prevention, "SARS Update," CDC Telebriefing Transcript, March 29, 2003.

p. 3 *soon identified the infective agent as related to a family of viruses that cause the common cold*: Centers for Disease Control and Prevention, "CDC Lab Analysis Suggests New Coronavirus May Cause SARS," press release, March 24, 2003; World Health Organization, "Severe Acute Respiratory Syndrome (SARS) Multi-Country Outbreak," Update 7, March 22, 2003.

p. 3 *the coronavirus probably came from a nonhuman animal*: British Columbia Cancer Agency, "2003 News—2003/04/12: Genome Sciences Centre Sequences SARS Associated Corona Virus," April 4, 2003; Centers for Disease Control and Prevention, "SARS: Genetic Sequencing of Coronavirus," CDC Telebriefing Transcript, April 14, 2003.

p. 3 *nearly 75 percent of new human diseases . . . are carried by wild or domestic animals*: P. Daszak, personal communication, April 2003; P. Daszak, A. A. Cunningham, and A. D. Hyatt, "Emerging Infectious Diseases of Wildlife: Threats to Biodiversity and Human Health," *Science* 287, no. 5452 (2000): 443–449.

p. 3 *smallpox . . . the common cold, from horses*: T. McMichael, *Human Frontiers, Environments, and Disease: Past Patterns, Uncertain Futures* (Cambridge, England: Cambridge University Press, 2001).

p. 3 *almost no way to eradicate it*: G. Kolata, "Now That SARS Has Arrived, Will It Ever Leave?" *New York Times*, April 27, 2003. Virologist and Nobel laureate Frederick C. Robbins is quoted as saying, "If you have an animal reservoir, unless you eradicate it in the animal, you can't eradicate it."

p. 4 *monkeys infected with Ebola virus were imported into Virginia*: Centers for Disease Control and Prevention, "Ebola Virus Infection in Imported Primates—Virginia, 1989," *Morbidity and Mortality Weekly Report* 38, no. 48 (1989): 831–832, 837–838.

p. 4 *"close the book on infectious diseases"*: McMichael, *Human Frontiers*, 88.

p. 4 *"diseases that seemed to be subdued . . . are fighting back with renewed ferocity"*: World Health Organization, *The World Health Report 1996: Fighting Disease, Fostering Development* (Geneva: World Health Organization, 1996), 1.

p. 4 *"exacerbate social and political instability in key countries and regions"*: J. C. Gannon, *The Global Infectious Disease Threat and Its Implications for the United States,* NIE 99-17D (Washington, D.C.: Central Intelligence Agency, 2000), 5.

p. 5 *residents of Chagugang . . . rioted:* E. Eckholm, "Fear: SARS Is the Spark for a Riot in China," *New York Times,* April 28, 2003.

p. 5 *kills nearly 2 million people annually. Half the victims are children under five years of age:* P. Martens and L. Hall, "Malaria on the Move: Human Population Movement and Malaria Transmission," *Emerging Infectious Diseases* 6, no. 2 (2000): 103–109.

p. 5 *resistant to chloroquine, a mainstay of malaria treatment:* X.Z. Su et al., "Complex Polymorphisms in an 330 kDa Protein Are Linked to Chloroquine-Resistant *P. falciparum* in Southeast Asia and Africa," *Cell* 91 (1997): 593–603.

p. 5 *malaria had been found in both humans and mosquitoes in an American community:* N. Wade, "Two Cases of Malaria Are Acquired in U.S., a Rarity," *New York Times,* September 7, 2002; "Virginia Mosquitoes Found with Malaria," *New York Times,* September 29, 2002.

p. 5 *increase in malaria has been linked to a warming global climate:* S. I. Hay et al., "Climate Change and the Resurgence of

Malaria in the East African Highlands," *Nature* 415, no. 6874 (2002): 905–909.

p. 5 *Maui, Hawai'i, reported its first case of dengue fever in more than fifty years:* Health Canada, Population and Public Health Branch, "Travel Health Advisory: Dengue Fever in Hawaii," April 18, 2002.

p. 6 *contributed to a 60 percent rise in fatal infections in the United States alone:* R. W. Pinner et al., "Trends in Infectious Diseases Mortality in the United States," *Journal of the American Medical Association* 275 (1996): 189–193.

p. 6 *A host of lesser-known new diseases and infectious agents also contribute to WHO's list:* McMichael, *Human Frontiers;* P. R. Epstein, "Climate Change and Emerging Infectious Diseases," *Microbes and Infection* 3, no. 9 (2001): 747–754; World Health Organization, "Emerging and Re-emerging Infectious Disease," Fact Sheet no. 97, revised August 1998.

p. 7 *Plagues are striking a wide range of other species:* Daszak, Cunningham, and Hyatt, "Emerging Infectious Diseases of Wildlife."

p. 7 *a brain-destroying affliction . . . is spreading among wild deer and elk in the western United States:* Ibid.

p. 8 *many new epidemics linked to environmental change . . . might rightfully be called "ecodemics":* In coining the term "ecodemics" to describe these outbreaks, I am not attempting to replace the word "epidemic" but merely emphasizing the ecological origins of many new diseases.

p. 8 *"Show me almost any new infectious disease":* Daszak, personal communication. April 2002.

p. 8 *major, extended waves of epidemics:* McNeill, *Plagues and Peoples;* McMichael, *Human Frontiers.*

p. 8 *close human contact with cattle and other livestock gave microbes a new bridge for jumping to humans:* McMichael, *Human*

Frontiers. McMichael writes (p. 101): "Smallpox arose via a mutant pox virus from cattle. Measles is thought to have come from the virus that causes distemper in dogs . . . the common cold from horses, and so on."

p. 9 *global exploration then ushered in a third phase of epidemics:* McMichael, *Human Frontiers.*

p. 9 *a fourth phase of epidemics:* Ibid.

p. 12 *Nipah virus:* World Health Organization, "Nipah Virus Fact Sheet," Fact Sheet no. 262, September 2001; K. B. Chua et al., "Nipah Virus: A Recently Emergent Deadly Paramyxovirus," *Science* 288, no. 5470 (2000): 1432–1435; P. Daszak, A. A. Cunningham, and A. D. Hyatt, "Anthropogenic Environmental Change and the Emergence of Infectious Diseases in Wildlife," *Acta Tropica* 78, no. 2 (2001): 103–116.

p. 13 *Pfiesteria piscicida:* J. Burkholder and H. B. Glasgow, "History of Toxic *Pfiesteria* in North Carolina Estuaries from 1991 to the Present," *BioScience* 51, no. 10 (2001): 827–841.

p. 14 *associations between ecological change and the emergence of a human epidemic . . . seen in Lyme disease:* R. S. Ostfeld, "The Ecology of Lyme-Disease Risk," *American Scientist* 85 (1997): 338–346.

p. 14 *Warmer temperatures, on average, . . . may have helped the disease-carrying tick expand its range:* M. L. Wilson, "Distribution and Abundance of *Ixodes scapularis* (Acari: Ixodidae) in North America: Ecological Processes and Spatial Analysis," *Journal of Medical Entomology* 35, no. 4 (1998): 446–457.

p. 16 *"Whether we like it or not, we are caught in the food chain":* Quoted in S. S. Morse, ed., *Emerging Viruses* (New York: Oxford University Press, 1993), 36.

Chapter 1. The Dark Side of Progress: Mad Cow Disease

p. 19 *"majestic herds of cattle":* W. Wordsworth, "Memorials of a

Tour on the Continent, 1820," no. 35, "After Landing—the Valley of Dover, November 1820," in *The Complete Poetical Works* (London: Macmillan, 1888).

p. 23 *"spongiform encephalopathy" on the necropsy form:* Copy of necropsy form signed by C. Richardson and dated September 19, 1985.

p. 23 *Wells confirmed Richardson's diagnosis:* N. Phillips, J. Bridgeman, and M. Ferguson-Smith, "The Identification of a New Disease in Cattle," chap. 1 in Phillips, Bridgeman, and Ferguson-Smith, *The BSE Inquiry,* vol. 3 (London: House of Commons, 2000), sec. 1.7; Phillips, Bridgeman, and Ferguson- Smith, "Statement No. 69: Carol Richardson," in *BSE Inquiry.*

p. 23 *"a novel progressive spongiform encephalopathy in cattle":* G. A. H. Wells et al., "A Novel Progressive Spongiform Encephalopathy in Cattle," *Veterinary Record* 121 (1987): 419–420.

p. 23 *the first-ever documented case of mad cow disease:* Phillips, Bridgeman, and Ferguson-Smith, "Identification of a New Disease in Cattle."

p. 24 *The first human TSE, Creutzfeldt-Jakob disease:* P. Brown and R. Bradley, "1755 and All That: A Historical Primer of Transmissible Spongiform Encephalopathy," *British Medical Journal* 317, no. 7174 (1998): 1688–1692.

p. 24 *another human TSE, kuru, was identified in Papua New Guinea:* Ibid.

p. 24 *By 1988 more than 2,000 cows had been stricken:* World Health Organization, "Bovine Spongiform Encephalopathy (BSE)," Fact Sheet no. 113, revised June 2001; Department for Environment, Food and Rural Affairs (London), *Bovine Spongiform Encephalopathy in Great Britain: A Progress Report —December 2001* (London: Department for Environment, Food and Rural Affairs, 2001).

p. 24 *almost 1,000 new cases . . . every week:* Centers for Disease

Control and Prevention, "New Variant CJD: Fact Sheet," April 18, 2002.

p. 24 *"vets have no cure"*: D. Brown, "Incurable Disease Wiping Out Dairy Cows," *Sunday Telegraph* (London), October 25, 1987.

p. 25 *woman . . . diagnosed with Creutzfeldt-Jakob disease*: Phillips, Bridgeman, and Ferguson-Smith, *BSE Inquiry;* "Emergence of Variant CJD," chap. 5 in Phillips, Bridgeman, and Ferguson-Smith, *BSE Inquiry,* vol. 8, *Variant CJD (vCJD),* sec. 5.3.

p. 25 *according to one study*: Phillips, Bridgeman, and Ferguson-Smith, *BSE Inquiry;* "History of CJD Surveillance up to 1990," chap. 2 in Phillips, Bridgeman, and Ferguson-Smith, *BSE Inquiry,* vol. 8, *Variant CJD (vCJD),* sec. 2.4.

p. 25 *average age of victims . . . was fifty-seven:* Phillips, Bridgeman, and Ferguson-Smith, *BSE Inquiry;* "History of CJD Surveillance up to 1990," chap. 2 in Phillips, Bridgeman, and Ferguson-Smith, *BSE Inquiry,* vol. 8, *Variant CJD (vCJD),* sec. 2.4.

p. 25 *CJD in the young:* Phillips, Bridgeman, and Ferguson-Smith, "Emergence of Variant CJD," chap. 5 in *BSE Inquiry,* vol. 8, *Variant CJD (vCJD),* sec. 5.28.

p. 25 *Conventional wisdom held:* S. B. Prusiner, "The Prion Diseases," *Scientific American* (January 1995): 48–57.

p. 25 *woman had been associated with a farm:* Phillips, Bridgeman, and Ferguson-Smith, "Emergence of Variant CJD," chap. 5 in *BSE Inquiry,* vol. 8, *Variant CJD (vCJD),* sec. 5.3.

p. 25 *"the risk of Bovine Spongiform Encephalopathy to humans is remote"*: Phillips, Bridgeman, and Ferguson-Smith, "The Southwood Working Party, 1988–1989," chap. 4 in *BSE Inquiry,* vol. 2, *The Link between BSE and vCJD,* sec. 4.1. This conclusion is stated in various ways throughout the *BSE Inquiry.* See also, for example, Phillips, Bridgeman, and Ferguson-Smith, "History of CJD Surveillance up to 1990," chap. 2 in *BSE Inquiry,* vol. 8, *Variant CJD (vCJD),* sec. 2.26.

. . . .

p. 25 *a sixty-one-year-old dairy farmer:* Phillips, Bridgeman, and Ferguson-Smith, "Emergence of Variant CJD," chap. 5 in *BSE Inquiry,* vol. 8, *Variant CJD (vCJD),* sec. 5.7; S. J. Sawcer et al., "Creutzfeldt-Jakob Disease in an Individual Occupationally Exposed to BSE," *Lancet* 341 (1993): 642.

p. 25 *"first report of CJD in an individual with direct occupational contact":* Sawcer et al., "Creutzfeldt-Jakob Disease in an Individual."

p. 26 *dismissed the notion of a link between mad cow disease and CJD:* Quoted in D. Brown, "Farmer Dies of Rare Brain Disease after BSE Hits Herd," *Daily Telegraph* (London), March 9, 1993.

p. 26 *"no evidence . . . of BSE passing from animals to humans":* Quoted in "Special Report: What's Wrong with Our Food? How the Death of Cow 133 Started a Tragic Chain of Events," *Observer* (London), October 1, 2000.

p. 26 *had vouched for the safety of beef:* BBC-TV, "John Gummer and His Daughter," May 1990.

p. 26 *"there is no need for people to be worried":* Ibid.

p. 26 *There had been three cases of BSE on his farm:* P. T. G. Davies, S. Jahfar, and I. T. Ferguson, "Creutzfeldt-Jakob Disease in Individual Occupationally Exposed to BSE," *Lancet* 342 (1993): 680; Phillips, Bridgeman, and Ferguson-Smith, "Emergence of Variant CJD," chap. 5 in *BSE Inquiry,* vol. 8, *Variant CJD (vCJD),* secs. 5.16–5.24; J. Hope, "'Mad Cow' Farmer Dies," *Daily Mail* (London), August 12, 1993; K. Perry, "Mad Cow Fear as Second Farmer Dies," *Today* (London), August 12, 1993.

p. 27 *a third dairy farmer:* P. E. M. Smith, M. Z. Zeidler, and J. W. Ironside, "Creutzfeldt-Jakob Disease in a Dairy Farmer," *Lancet* 346 (1995): 898; "Illness Link with Mad Cow Disease?" *Cornishman* (Cornwall), December 1, 1994; Phillips, Bridgeman, and Ferguson-Smith, "Emergence of Variant

CJD," chap. 5 in *BSE Inquiry,* vol. 8, *Variant CJD (vCJD),* secs. 5.16–5.33.

p. 27 *did not require "the Government to revise the measures":* Phillips, Bridgeman, and Ferguson-Smith, "Emergence of Variant CJD," chap. 4 in *BSE Inquiry,* vol. 11, *Annex 2 to Chapter 4: SEAC Meetings* (see the heading "Special Meeting 13/1/95").

p. 27 *should a fourth case arise, the tide of probability would turn:* Phillips, Bridgeman, and Ferguson-Smith, "Emergence of Variant CJD," chap. 5 in *BSE Inquiry,* vol. 8, *Variant CJD (vCJD),* sec. 5.39.

p. 27 a *fourth ill farmer came to light:* Ibid., sec. 5.66.

p. 27 *fifteen-year-old . . . came down with CJD:* Phillips, Bridgeman, and Ferguson-Smith, "Emergence of Variant CJD," chap. 5 in *BSE Inquiry,* vol. 8, *Variant CJD (vCJD),* sec. 5.25; World Health Organization, "Possible Creutzfeldt-Jakob Disease in an Adolescent," *Weekly Epidemiological Record* 15 (1994): 105–106.

p. 28 *beef burger was Vickie's favorite food:* A. Watkins, "Today Investigation," *Today* (London), January 13, 1994.

p. 28 *"no evidence whatever that BSE causes CJD":* Phillips, Bridgeman, and Ferguson-Smith, "Emergence of Variant CJD," chap. 5 in *BSE Inquiry,* vol. 8, *Variant CJD (vCJD),* sec. 5.29; see also footnote 220, citing "Statement of CMO on CJD," Department of Health press release, January 26, 1994 (YB94/1.26/3.1).

p. 28 *coma that lasted four and a half years:* Phillips, Bridgeman, and Ferguson-Smith, "Statement No. 208: Beryl Rimmer," in *BSE Inquiry.*

p. 28 *"ingestion of the infective agent may be one natural mode":* M. Kamin and B. M. Patten, "Creutzfeldt-Jakob Disease: Possible Transmission to Humans by Consumption of Wild Animal Brains," *American Journal of Medicine* 76 (1984): 142–145.

p. 29 *minimized heredity or direct contact:* J. R. Berger, E. Weis-

· · · · · · · ·

man, and B. Weisman, "Creutzfeldt-Jakob Disease and Eating Squirrel Brains," *Lancet* 350, no. 9078 (1997): 642.

p. 29 *Analysis of cells from the man's and the cat's brains showed remarkably similar abnormalities:* G. Zanusso et al., "Simultaneous Occurrence of Spongiform Encephalopathy in a Man and His Cat in Italy," *Lancet* 352, no. 9134 (1998): 1116–1117.

p. 30 *"possibility that [these cases were] causally linked to BSE":* R. G. Will et al., "A New Variant of Creutzfeldt-Jakob Disease in the UK," *Lancet* 347, no. 9006 (1996): 921–925.

p. 30 *indistinguishable from the agent that caused BSE:* A. F. Hill et al., "The Same Prion Strain Causes vCJD and BSE," *Nature* 389 (1997): 448–450.

p. 31 *animals "lie down . . . and finally become lame":* Brown and Bradley, "1755 and All That," 1688.

p. 31 *"malady of madness and convulsions":* H. B. Parry, *Scrapie Disease in Sheep: Historical, Clinical, Epidemiological, Pathological, and Practical Aspects of the Natural Disease* (New York: Academic Press, 1983).

p. 31 *origins remained a mystery:* Brown and Bradley, "1755 and All That."

p. 31 *scrapie agent remained infectious:* Centers for Disease Control and Prevention and National Institutes of Health, "Prions," in *Biosafety in Biomedical and Microbiological Laboratories,* edited by J. Y. Richmond and R. W. McKinney (Washington, D.C.: U.S. Government Printing Office, 1999), sec. VII-D.

p. 32 *a nonliving infectious agent:* Brown and Bradley, "1755 and All That."

p. 32 *disease that was killing the Foré people:* Ibid.

p. 32 *brains of kuru victims looked a lot like those of CJD victims:* Ibid.; R. B. Wickner et al., *Prions in Yeast Are Protein Genes: Inherited Amyloidosis* (Bethesda, Md.: National Institutes of Health, National Institute of Diabetes and Digestive and Kidney Diseases, 2002); Karolinska Institutet, "The 1997 Nobel

Prize in Physiology or Medicine," press release, October 6, 1997.

p. 32 *three brain-wasting diseases came to largely define TSEs:* Brown and Bradley, "1755 and All That."

p. 32 *1976 Nobel Prize in Physiology or Medicine:* Karolinska Institutet, "The 1976 Nobel Prize in Physiology or Medicine," press release, October 14, 1976.

p. 33 *prions, do not reproduce:* S. B. Prusiner, "The Prion Disease."

p. 33 *Scrapie-infected sheep remained the top suspect:* J. Hope et al., "Molecular Analysis of Ovine Prion Protein Identifies Similarities between BSE and an Experimental Isolate of Natural Scrapie, CH1641," *Journal of General Virology* 80 (1999): 1–4.

p. 33 *not a single documented case of a cow becoming sick from scrapie:* M. Balter, "On the Hunt for a Wolf in Sheep's Clothing," *Science* 287, no. 5460 (2000): 1906–1908.

p. 33 *Nor . . . was a single case documented of a person becoming sick from the sheep disease:* Ibid.

p. 34 *Perhaps a protein in a cow's brain had randomly mutated:* C. A. Donnelly, Department of Infectious Disease Epidemiology, Imperial College London, personal communication, November 26, 2002.

p. 35 *renderers in France:* European Commission, *Report of the European Union with Regard to the Implementation of Council Directive 90/667/EEC and Commission Decision 91/516/EEC concerning the Use of Prohibited Ingredients in the Animal Feedingstuff (19–20 August 1999)*, ref. no. XXIV/1234/99 (Brussels: European Union, 1999). The report states: "Certain plants in the French rendering industry have used for years prohibited substances such as sludge from the biological treatment of the waste water or water from septic tanks from their own establishments or, possibly, from their suppliers."

p. 37 *MBM. . . . is added to animal feed:* J. Lederberg, R. E. Shope, and S. C. Oaks Jr., eds., *Emerging Infections: Microbial Threats* to Health in the United States (Washington, D.C.: National Academy Press, 1992).

p. 37 *dramatic increase in the number of sheep in the United Kingdom:* Department for Environment, Food and Rural Affairs (London), "Sheep and Lambs; Mutton and Lamb," table 5.14 in *Agriculture in the United Kingdom* (London: Department for Environment, Food and Rural Affairs, 2002).

p. 38 *BSE rapidly spread . . . via feed containing meat and bone meal of infected animals:* J. W. Wilesmith et al., "Bovine Spongiform Encephalopathy: Epidemiological Studies," *Veterinary Record* 123 (1988): 638–644.

p. 38 *"a recipe for disaster":* Phillips, Bridgeman, and Ferguson-Smith, "Executive Summary of the Report of the Inquiry," chap. 1 in *BSE Inquiry,* vol. 1; see "Key Conclusions" in *Findings and Conclusions.*

p. 38 *1988 ban on the feeding of recycled animal protein:* Donnelly, personal communication.

p. 39 *an estimated 1 million cows had been infected:* Ibid.

p. 39 *cases had been confirmed in more than 35,000 herds:* Centers for Disease Control and Prevention, "Questions and Answers Regarding Bovine Spongiform Encephalopathy (BSE) and Creutzfeldt-Jakob Disease (CJD)," on-line at http://www.cdc. gov/ncidod/diseases/cjd/bse_cjd_qa.htm (last reviewed June 3, 2003).

p. 39 *in Belgium, Denmark, Switzerland, Italy, Greece, Germany, France, the Netherlands, Portugal, Ireland, and Spain:* Office International des Épizooties, "Number of Reported Cases of Bovine Spongiform Encephalopathy (BSE) Worldwide as of 04.03.2003" (Paris: Office International des Épizooties, 2003); Centers for Disease Control and Prevention, "Update 2002: Bovine Spongiform Encephalopathy and Variant Creutzfeldt-Jakob Disease," on-line at http://www.cdc.gov/ncidod/diseases/cjd/cjd.htm (last reviewed June 5, 2005).

p. 39 *125 cases of the human form of mad cow disease:* Centers for Disease Control and Prevention, "New Variant CJD: Fact Sheet."

p. 39 *history of exposure . . . where the disease was occurring in cattle:* Ibid.

p. 39 *banned the practice of feeding animal by-products to cattle:* Donnelly, personal communication.

p. 39 *"highly unlikely that a person would contract vCJD . . . in the United States":* Food and Drug Administration, Center for Food Safety and Applied Nutrition, "Consumer Questions and Answers about BSE," March 2001.

p. 39 *the disease came perilously close to the United States:* Health Canada, "BSE Disease Investigation in Alberta," news release, May 20, 2003; S. Blakeslee, "Mad Cow Disease Is Found in Canada; U.S. Imposes a Ban," *New York Times,* May 20, 2003.

p. 40 *Letter to Christopher Melani:* Photocopy of letter from Michael W. Miller to Christopher Melani, December 20, 1997.

p. 41 *"What's done is done":* Quoted in "Wasting Away in the West," CBS Evening News, March 13, 2001.

p. 42 *by 2001 the disease had spread to Nebraska:* U.S. Department of Agriculture, Animal and Plant Health Inspection Service, Veterinary Services, "APHIS Factsheet," October 2001.

p. 42 *three young venison eaters:* E. D. Belay et al., "Creutzfeldt-Jakob Disease in Unusually Young Patients Who Consumed Venison," *Archives of Neurology* 58, no. 10 (2001): 1673–1678.

p. 43 *five reported vCJD cases:* Centers for Disease Control and Prevention, "New Variant CJD: Fact Sheet."

p. 43 *more than worrisome:* Ibid.

p. 43 *victims' median age was sixty-eight:* Ibid.; R. V. Gibbons et al., "Creutzfeldt-Jakob Disease in the United States: 1979–1998," *Journal of the American Medical Association* 284, no. 18 (2000): 2322–2323.

p. 43 *"more likely to get run over by a Winnebago"*: T. Thorne, personal communication, July 2002.

p. 44 *"the whole issue of prion disease . . . has to be confronted seriously"*: S. Rampton, "What about Mad Deer Disease?" *E Magazine* 12, no. 4 (July–August 2001).

p. 44 *Brains of young victims*: P. Yam, "Shoot this deer," *Scientific American,* June 2003: 38–43.

p. 44 *chronic wasting disease . . . is as infectious to human tissue as mad cow disease*: G. J. Raymond et al., "Evidence of a Molecular Barrier Limiting Susceptibility of Humans, Cattle, and Sheep to Chronic Wasting Disease," *EMBO Journal* 19 (2000): 4425–4430.

p. 44 *a cow was experimentally infected*: C. Q. Choi, "Chronic Wasting Studies Announced," *Scientist,* November 8, 2002.

p. 44 *new studies to determine CWD's contagiousness*: Ibid.

p. 45 *officials began shooting deer from helicopters*: S. Blakeslee, "Clues to Mad Cow Disease Emerge in Study of Mutant Proteins," *New York Times*, May 23, 2000; S. Blakeslee, "Weighing 'Mad Cow' Risks in American Deer and Elk," *New York Times*, February 23, 1999.

Chapter 2. A Chimp Called Amandine: HIV/AIDS

p. 47 *"the earliest beginnings of the world"*: J. Conrad, "Heart of Darkness," in *The Portable Conrad,* edited by M. D. Zabel (New York: Viking Press, 1950), 536.

p. 49 *a new immunosuppressive disease . . . had emerged*: M. S. Gottlieb et al., "Pneumocystis Pneumonia—Los Angeles," *Morbidity and Mortality Weekly Report* 30, no. 21 (1981): 250–252.

p. 49 *the name "acquired immunodeficiency syndrome"—AIDS—was coined*: Centers for Disease Control and Prevention, "Where Did HIV Come From?" on-line at http://www.cdc.gov/hiv/pubs/faq/faq3.htm (last reviewed June 2, 2003).

p. 50 *immune-ravaged patients in Europe*: M. D. Grmek, *History of*

AIDS: Emergence and Origin of a Modern Pandemic (Princeton, N.J.: Princeton University Press, 1990).

p. 50 *researchers in the United States discovered a frozen blood sample:* M. Balter, "Virus from 1959 Sample Marks Early Years of HIV," *Science* 279, no. 5352 (1998): 801; A. G. Motulsky, J. Vandepitte, and G. R. Fraser, "Population Genetic Studies in the Congo: I. Glucose-6-Phosphate Dehydrogenase Deficiency, Hemoglobin S, and Malaria," *American Journal of Human Genetics* 18, no. 6 (1966): 514–537.

p. 50 *man was from Leopoldville, Belgian Congo:* T. Zhu et al., "An African HIV-1 Sequence from 1959 and Implications for the Origin of the Epidemic," *Nature* 391, no. 6667 (1998): 594–597.

p. 50 *earliest documented case of HIV-1 infection:* Balter, "Virus from 1959 Sample."

p. 53 *an SIV . . . almost indistinguishable from HIV-2:* V. M. Hirsch et al., "An African Primate Lentivirus (SIVsm) Closely Related to HIV-2," *Nature* 339, no. 6223 (1989): 389–392. R. V. Gilden et al. "HTLV-III Antibody in a Breeding Chimpanzee Not Experimentally exposed to the virus," *The Lancet,* March 22, 1986: 94–95.

p. 54 *died from complications of childbirth:* "Pathology Worksheet (Necropsy): Animal I.D. Number 205" (Alamogordo: New Mexico State University, Primate Research Institute, 1985).

p. 55 *They named her Amandine:* Martine Peters, personal communication, June 2002.

p. 55 *a series of seizures:* J. C. Vié et al., "Megaloblastic Anemia in a Handreared Chimpanzee," *Laboratory Animal Science* 39, no. 6 (1989): 613–615.

p. 56 *the test came back positive:* M. Peeters et al., "Isolation and Partial Characterization of an HIV-Related Virus Occurring Naturally in Chimpanzees in Gabon," *AIDS* 3, no. 10 (1989): 625–630.

p. 57 *Macolamapoye*: M. Peeters, personal communication, June 2002.

p. 57 *Delaporte, a physician, was plenty worried*: E. Delaporte, personal communication, June 2002.

p. 58 *virus fragments found in chimps were indeed related to HIV-1*: T. Huet et al., "Genetic Organization of a Chimpanzee Lentivirus Related to HIV-1," *Nature* 345, no. 6273 (1990): 356–359.

p. 61 *"everyone's nightmare"*: L. K. Altman, "H.I.V. Linked to a Subspecies of Chimpanzee," *New York Times*, February 1, 1999.

p. 61 *"one of their main sources of food is bushmeat"*: Peeters, personal communication; see also M. C. Peeters et al., "Risk to Human Health from a Plethora of Simian Immunodeficiency Viruses in Primate Bushmeat," *Emerging Infectious Diseases* 8, no. 5 (2002): 451–457.

Chapter 3. The Travels of Antibiotic Resistance: Salmonella DT104

p. 64 *first, in 1996, struck nineteen schoolchildren*: R. G. Villar et al., "Investigation of Multi-Resistant *Salmonella* Serotype *typhimurium* DT104 Infections Linked to Raw-Milk Cheese in Washington State," *Journal of the American Medical Association* 281, no. 19 (1999): 1811–1816.

p. 65 *eating unpasteurized Mexican-style soft cheese*: S. H. Cody et al., "Two Outbreaks of Multi-Resistant *Salmonella* Serotype *typhimurium* DT104 Infections Linked to Raw-Milk Cheese in Northern California," *Journal of the American Medical Association* 281, no. 19 (1999): 1805–1810.

p. 65 *pathologist Theobald Smith*: C. E. Dolman, "Theobald Smith, 1859–1934: A Fiftieth Anniversary Tribute," *American Society of Microbiology News* 50, no. 12 (1984): 577–580.

p. 65 *genus of bacteria that live in the intestines of many species*: R. K. Robinson, C. A. Batt, and P. Patel, eds., *Encyclopedia of*

Food Microbiology, vol. 3, *Salmonella* (London: Academic Press, 2000).

p. 69 *by 1963 type 29 had become resistant to two antibiotics:* E. S. Anderson and M. J. Lewis, "Drug Resistance and Its Transfer in *Salmonella typhimurium,*" *Nature* 206, no. 4984 (1965): 579–583.

p. 70 *rare forms had armed themselves against seven antibiotics:* Ibid.; E. S. Anderson, "Drug Resistance in *Salmonella typhimurium* and Its Implications," *British Medical Journal* 3 (1968): 333–339.

p. 70 *Of some 500 confirmed human cases . . . 6 were fatal:* Anderson, "Drug Resistance in *Salmonella typhimurium* and Its Implications."

p. 70 *the dealer . . . apparently committed suicide:* L. Ward, head of the salmonella research unit, Public Health Laboratory Service, personal communication, June 2002.

p. 70 *outbreaks were "almost entirely of bovine origin":* Anderson, "Drug Resistance in *Salmonella typhimurium* and Its Implications."

p. 70 *"the time has clearly come for a re-examination":* Anderson and Lewis, "Drug Resistance and Its Transfer in *Salmonella typhimurium,*" p. 583.

p. 70 *use of antibiotics to make animals grow faster "should be abolished altogether":* Editorial, "A Bitter Reckoning," *New Scientist* (January 4, 1968): 14–15.

p. 70 *"Unless drastic measures are taken":* J. Bower, "The Farm Drugs Scandal," *Ecologist* 1 (1970): 10–15.

p. 71 *"there is ample data now in the literature":* FDA report as cited by T. H. Jukes in "Public Health Significance of Feeding Low Levels of Antibiotics to Animals," *Advances in Applied Microbiology* 16 (1973): 1–29; quote, p. 19.

p. 71 *indiscriminate antibiotic use "favors the . . . development of single- and multiple-antibiotic-resistant bacteria":* C. D. Van

. . . .

Houweling, Food and Drug Administration, press briefing statement, January 31, 1972.

p. 71 *licenses for use of antibiotics as growth promotants should be revoked*: Food and Drug Administration, *Antibiotics in Animal Feeds: Information for Consumers* (Rockville, Md.: Food and Drug Administration, Center for Veterinary Medicine, 1993); M. Mellon, "Antibiotic Resistance: Causes and Cures" (speech given to the National Press Club, Washington, D.C., June 4, 1999).

p. 71 *"a cult of food quackery"*: Jukes, "Public Health Significance."

p. 71 *cited as evidence of the cult*: Ibid.

p. 71 *advocated that antibiotics be routinely used in some human food*: Ibid.

p. 72 *Antibiotics, he believed, could compensate for malnourishment*: Ibid.

p. 72 *incidence of salmonella type 29 declined*: E. J. Threlfall et al., "The Emergence and Spread of Antibiotic Resistance in Food-Borne Bacteria in the United Kingdom," *APUA Newsletter* (published by the Alliance for the Prudent Use of Antibiotics) 17, no. 4 (1999): 1–7.

p. 73 *spread to Cambridgeshire and Yorkshire*: Editorial, "Why Has Swann Failed?" *British Medical Journal* 1, no. 6225 (1980): 1195–1196.

p. 73 *"man who threw himself out of the Empire State Building"*: R. Young et al., *The Use and Misuse of Antibiotics in UK Agriculture. Part 2: Antibiotic Resistance and Human Health* (Bristol, England: Soil Association, 1999).

p. 73 *pharmaceuticals that carried a higher profit margin*: Ibid.

p. 74 *half a dozen weighty scientific evaluations*: R. A. Stallones, "Epidemiology and Public Policy: Pro- and Anti-Biotic," *American Journal of Epidemiology* 115, no. 4 (1982): 485–491.

p. 74 *outspoken advocate for unlimited antibiotic use*: P. B. Lieberman and M. G. Wootan, *Protecting the Crown Jewels of Medi-*

cine: A Strategic Plan to Preserve the Effectiveness of Antibiotics (Washington, D.C.: Center for Science in the Public Interest, 1998).

p. 74 *"If the decision were mine"*: Stallones, "Epidemiology and Public Policy," 490.

p. 75 *solved the case without using food as the witness*: T. F. O'Brien et al., "Molecular Epidemiology of Antibiotic Resistance in *Salmonella* from Animals and Human Beings in the United States," *New England Journal of Medicine* 307 (1982): 1–6.

p. 75 *issue "certainly has been studied sufficiently"*: B. Keller, "Ties to Human Illness Revive Move to Ban Medicated Feed," *New York Times*, September 16, 1984.

p. 75 *Britain's ban of certain antibiotics . . . suggested exactly the opposite*: Ibid.

p. 75 *budget . . . was in the hands of the same appropriations subcommittee*: Ibid.

p. 76 *Bactrim was another option*: F. J. Angulo et al., "Origins and Consequences of Antimicrobial-Resistant Nontyphoidal *Salmonella*: Implications for the Use of Fluoroquinolones in Food Animals," *Microbial Drug Resistance* 6, no. 1 (2000): 77–83; M. Vijups, personal communications, July 2002.

p. 77 *it struck seven people in Airdrie, Scotland*: Threlfall et al., "Emergence and Spread of Antibiotic Resistance"; R. Davies, *Zoonose-Nyt* 8, no. 1 (2001).

p. 79 *"bacteria that lived on farmed fish in Southeast Asia"*: E. H. Kim and T. Aoki, "Drug Resistance and Broad Geographical Distribution of Identical R Plasmids of *Pasteurella piscicida* Isolated from Cultured Yellowtail in Japan," *Microbiology and Immunology* 37, no. 2 (1993): 103–109.

p. 80 *traced to fish meal from Peru*: G. M. Clark, A. F. Kaufmann, and E. J. Gangrosa, "Epidemiology of an International Outbreak of *Salmonella agona*," *Lancet* (1973): 490–493.

p. 82 *hailed as a breakthrough treatment for many infections*:

Lieberman and Wootan, *Protecting the Crown Jewels of Medicine*; S. J. Olsen et al., "A Nosocomial Outbreak of Fluoroquinolone-Resistant *Salmonella* Infection," *New England Journal of Medicine* 344 (2001): 1572–1579.

p. 82 *began to show resistance to fluoroquinolones:* H. P. Endz et al., "Quinolone Resistance in *Campylobacter* Isolated from Man and Poultry Following the Introduction of Fluoroquinolones in Veterinary Medicine," *Journal of Antimicrobial Chemotherapy* 27 (1991): 199–208.

p. 83 *licensed fluoroquinolone for treating and preventing illness in turkeys and chickens:* Angulo et al., "Origins and Consequences."

p. 83 *Two years later, 16 percent . . . showed some resistance:* E. J. Threlfall, "Increasing Spectrum of Resistance in Multiresistant *Salmonella typhimurium*," *Lancet* 347, no. 9007 (1996): 1053–1054.

p. 83 *by 1996, fluoroquinolone-resistant . . . infections were sickening people:* Threlfall et al., "Emergence and Spread of Antibiotic Resistance."

p. 83 *full support of the CDC:* F. J. Angulo, personal communication.

p. 83 *in 1995 the FDA granted approval:* Angulo et al., "Origins and Consequences"; S. Rossiter et al., "Emerging Fluoroquinolone Resistance among Non-Typhoidal *Salmonella* in the United States: NARMS 1996–2000" (presentation given at International Conference on Emerging Infectious Diseases, Atlanta, Ga., March 26, 2002).

p. 83 *By 2000, 1.4 percent of salmonella infections showed some resistance:* Rossiter et al., "Emerging Fluoroquinolone Resistance."

p. 84 *meat products from supermarkets in the Minneapolis–St. Paul area:* S. E. Kirk, J. M. Besser, and W. C. Hedberg, "Quinolone-Resistant *Campylobacter jejuni* Infections in Minnesota, 1992–1998," *New England Journal of Medicine* 240, no. 20 (1999): 1525–1532.

p. 84 *strong evidence . . . that the use of fluoroquinolones in poultry posed a risk to human health:* Lieberman and Wootan, *Protecting the Crown Jewels of Medicine.*

p. 84 *this time the agency proposed revoking the license:* Food and Drug Administration, "FDA, CVM Proposes to Withdraw Poultry Fluoroquinolones Approval," *CVM Update,* October 26, 2000; "FDA Proposes to Ban Two Poultry Antibiotics," *Baltimore Sun,* October 28, 2000, 8A.

p. 84 *"The consensus is that there is no public health risk":* L. Fabregas, "Bayer, FDA Spar over Safety of Poultry Drug," *Pittsburgh Tribune-Review,* January 20, 2002.

p. 84 *"there is no scientific evidence":* Animal Health Institute, "Statement by Alexander S. Mathews, 'The Use of Antibiotics in Food-Producing Animals,'" press release, May 28, 1998.

p. 84 *"There is no clear documentation":* R. Carnevale et al., "Fluoroquinolone Resistance in *Salmonella:* A Web Discussion," *Clinical Infectious Diseases* 31 (2000): 128–130.

p. 85 *"no conclusive evidence":* R. Robinson, in *Lancaster New Era,* March 6, 2002.

p. 85 *"if we are what we eat, we're healthier if they're healthier":* S. Lerner, "Risky Chickens," *Village Voice,* November 28–December 4, 2001.

p. 85 *echoed the warnings from the 1960s:* "Bitter Reckoning."

p. 86 *wastewater contaminates streams, rivers and aquifers, and lakes and their shores:* D. W. Kolpin et al., "Pharmaceuticals, Hormones, and Other Organic Wastewater Contaminants in U.S. Streams, 1999–2000: A National Reconnaissance," *Environmental Science and Technology* 36, no. 6 (2002): 1202–1211; M. Meyer et al., *Occurrence of Antibiotics in Surface and Ground Water Near Confined Animal Feeding Operations and Waste Water Treatment Plants Using Radioimmunoassay and Liquid Chromatography/Electrospray Mass Spectrometry* (Raleigh, N.C.: U.S. Geological Survey, 2000).

p. 86 *wastewater treatment plants in Europe:* J. Raloff, "Drugged Waters: Does It Matter That Pharmaceuticals Are Turning Up in Water Supplies?" *Science News Online*, March 21, 1998.

p. 86 *two lakes in Switzerland:* R. Hirsch et al., "Occurrence of Antibiotics in the Aquatic Environment," *Science of the Total Environment* 225 (1999): 109–118.

p. 86 *sediments under fish farms:* Letter from Union of Concerned Scientists to the U.S. Environmental Protection Agency urging limits on vital antibiotics in factory farm effluent, August 3, 2000.

p. 86 *resistance to twelve different antibiotics:* D. G. White et al., "The Isolation of Antibiotic-Resistant *Salmonella* from Retail Ground Meats," *New England Journal of Medicine* 345, no. 16 (2001): 1147–1154.

Chapter 4. Of Old Growth and Arthritis: Lyme Disease

p. 89 *Harrison's purchase was but a tiny grove:* D. S. Wilcove, *The Condor's Shadow* (New York: Freeman, 1999). Although Wilcove does not speak of Harrison's purchase, he describes the vastness of the early eastern forests of which that purchase was a part.

p. 90 *farmland quilted the region:* E. B. Stryker, *Where the Trees Grow Tall* (Franklin Township, N.J.: Franklin Township Historical Society, 1963), 173; J. P. Snell, *History of Hunterdon and Somerset Counties New Jersey: With Illustrations and Biographical Sketches of Its Prominent Men and Pioneers* (Philadelphia: Everts & Peck, 1881), microfilm.

p. 90 *more than half of the . . . northeastern forests had been cut:* L. C. Irland, *The Northeast's Changing Forests* (Petersham, Mass.: Distributed by Harvard University Press for Harvard Forest, 1999).

p. 90 *several organizations . . . purchased the land:* E. Stiles, personal communication, June 2002.

p. 93   *today forests cover three-quarters of their historical range:*
Irland, *Northeast's Changing Forests.*

p. 93   *Farms flowed in . . . and then washed out:* Wilcove, *Condor's
Shadow.*

p. 94   *white-tailed deer were almost extirpated:* L. Ragonese, "Deer
Thrive as the Forests Sicken," *New Jersey Star-Ledger*, October
20, 1999. The article stated that white-tailed deer were "nearly
extinct in New Jersey 100 years ago."

p. 98   *refuge and breeding grounds for mice and chipmunks:* K. A.
Orloski et al., "Emergence of Lyme Disease in Hunterdon
County, New Jersey, 1993: A Case-Control Study of Risk Fac-
tors and Evaluation of Reporting Patterns," *American Journal
of Epidemiology* 147, no. 4 (1998): 391–397.

p. 99   *average temperature in nearby New Brunswick increased:* U.S.
Environmental Protection Agency, *Climate Change and New
Jersey* (Washington, D.C.: U.S. Environmental Protection
Agency, Office of Policy, Planning and Evaluation, 1997). The
report states that "over the last century, the average temperature in New Brunswick, New Jersey, has increased from 50.4
degrees F (1889–1918 average) to 52.2 degrees F (1966–1995
average), and precipitation in some locations in the state has
increased by 5–10%."

p. 99   *an explosion in tick populations:* M. L. Wilson, "Distribution
and Abundance of *Ixodes scapularis* (Acari: Ixodidae) in North
America: Ecological Processes and Spatial Analysis," *Journal of
Medical Entomology* 35, no. 4 (1998): 446–457; E. Lindgren
and R. Gustafson, "Tick-Borne Encephalitis in Sweden and
Climate Change," *Lancet* 358, no. 9275 (2001): 16–18. It should
be noted that some scientists attribute the spread of ticks pri-
marily to increasing deer populations.

p. 105   *a cascade effect:* R. S. Ostfeld et al., "Effects of Acorn Production and Mouse Abundance on Abundance and *Borrelia
burgdorferi* Infection Prevalence of Nymphal *Ixodes scapularis*

· · · · · · · ·

Ticks," *Vector Borne and Zoonotic Diseases* 1, no. 1 (2001): 55–63;
R. S. Ostfeld, "The Ecology of Lyme-Disease Risk," *American
Scientist* 85 (1997): 338–346.

p. 109 *third-highest number of Lyme disease cases.* R. Ostfeld, personal communication, November 2002.

p. 110 *areas with more species had fewer cases of Lyme disease per
capita*: R. S. Ostfeld and F. Keesing, "The Function of Biodiversity in the Ecology of Vector-Borne Zoonotic Diseases,"
Canadian Journal of Zoology 78 (2000): 2061–2078.

p. 111 *indigenous forest dwellers might . . . have developed immunity
to it*: J. A. Patz et al., "Effects of Environmental Change on
Emerging Parasitic Diseases," *International Journal for Parasitology* 30, no. 12–13 (2000): 1395–1405.

Chapter 5. A Spring to Die For: Hantavirus

p. 113 *"mysterious illness that killed two young Navajos"*: Excerpted
from Steve Sternberg, "An Outbreak of Pain," *USA Today*, July
2, 1998 (used by permission).

p. 116 *The Colorado Plateau*: H. D. Grissino-Mayer, University of
Tennessee, Knoxville, personal communication, May 2003; R.
D. D'Arrigo and G. C. Jacoby, "A 1000-Year Record of Winter
Precipitation from Northwestern New Mexico, USA: A
Reconstruction from Tree-Rings and Its Relation to El Niño
and the Southern Oscillation," *Holocene* 1 (1991): 95–101; H.
D. Grissino-Mayer, "A 2,129-Year Reconstruction of Precipitation for Northwestern New Mexico, USA," in *Tree Rings,
Environment, and Humanity: Proceedings of the International
Conference, Tucson, Arizona, 17–21 May, 1994*, edited by J. S.
Dean, D. M. Meko, and T. W. Swetnam (Tucson, Ariz.:
Radiocarbon, 1996), 191–204.

p. 116 *stretches across 130,000 square miles of southeastern Utah*: R.
Wheeler, "The Colorado Plateau Region," in *Wilderness at the*

Edge: A Citizen Proposal to Protect Utah's Canyons and Deserts (Salt Lake City: Utah Wilderness Coalition, 1990).

p. 118 *the extremes and duration of the heated Pacific appear to be something new:* K. Lewis and D. Hathaway, "Analysis of Paleo-Climate and Climate-Forcing Information for New Mexico and Implications for Modeling in the Middle Rio Grande Water Supply Study" (Boulder, Colo.: S. S. Papadopulos & Associates, 2001).

p. 118 *A powerful El Niño cycle began in 1991:* R. Merideth, *A Primer on Climatic Variability and Change in the Southwest* (Tucson: University of Arizona, Udall Center for Studies in Public Policy and Institute for the Study of Planet Earth, 2001).

p. 118 *quickly rising waters . . . sweeping a fifteen-year-old to his death:* Los Angeles County Office of Education and Department of Public Works, *No Way Out,* videotape produced by Nancy Rigg, 2000.

p. 118 *winds laden with Pacific moisture pummeled Las Vegas:* K. Rogers, "Damage from Worst Flood Parallels Growth," *Las Vegas Review-Journal,* July 9, 1999.

p. 119 *declared the normally arid state a flood disaster area:* Federal Emergency Management Agency, *Disaster Activity: January 1, 1992, to December 31, 1992* (Washington, D.C.: Federal Emergency Management Agency, 1992).

p. 119 *series of mild snowstorms interspersed with rain:* Merideth, *Primer on Climatic Variability.*

p. 119 *New Mexico was again declared a flood disaster area:* D. M. Engelthaler et al., "Climatic and Environmental Patterns Associated with Hantavirus Pulmonary Syndrome, Four Corners Region, United States," *Emerging Infectious Diseases* 5, no. 1 (1999): 87–94.

p. 119 *extensive roots quickly soaked up the water:* A. T. Carpenter and T. A. Murray, "Element Stewardship Abstract for *Bromus tectorum* L. (*Anisantha tectorum* (L.) Nevski)" (Arlington, Va.: Nature Conservancy, n.d.).

p. 119 *Snakeweed also burst forth:* Wyoming Agricultural Experiment Station, "Species Fact Sheet: Snakeweed Grasshopper, *Hesperotettix viridis* (Thomas)" (Laramie: Wyoming Agricultural Experiment Station, 1994).

p. 120 *a year after a devastating influenza epidemic struck:* A. Crosby, *America's Forgotten Pandemic: The Influenza of 1918* (Cambridge, England: Cambridge University Press, 1989); G. Bailey and R. G. Bailey, *A History of the Navajos: The Reservation Years* (Santa Fe, N.M.: School of American Research Press, 1986).

p. 120 *unusually heavy winter and spring rains:* Centers for Disease Control and Prevention, "Navajo Medical Traditions and HPS," 2000.

p. 122 *evergreen plant was used by some Native Americans:* A. B. Reagan, "Plants Used by the White Mountain Apache Indians of Arizona," *Wisconsin Archaeologist* 8 (1929): 143–161; A. F. Whiting, *Ethnobotany of the Hopi,* Bulletin no. 15 (1939; reprint, Flagstaff: Museum of Northern Arizona, 1966).

p. 122 *a cold medication . . . could loosen the patient's mucus:* S. A. Weber and P. D. Seaman, *Havasupai Habitat: A. F. Whiting's Ethnography of a Traditional Indian Culture* (Tucson: University of Arizona Press, 1985).

p. 125 *directly followed these increased densities of mice:* Centers for Disease Control and Prevention, "El Niño Special Report: Could El Niño Cause an Outbreak of Hantavirus Disease in the Southwestern United States?" 2000.

p. 125 *"increased winter and summer rain is associated with outbreaks of hantavirus":* J. N. Mills et al., "Long-Term Studies of Hantavirus Reservoir Populations in the Southwestern United States: A Synthesis," *Emerging Infectious Diseases* 5, no. 1 (1999): 135–142.

p. 125 *researchers from Johns Hopkins:* G. E. Glass et al., "Using Remotely Sensed Data to Identify Areas at Risk for Hantavirus Pulmonary Syndrome," *Emerging Infectious Diseases* 6, no. 3 (2000): 238–247.

p. 126 *"In beauty and harmony it shall so be finished":* G. Hausman, *Meditations with the Navajo: Prayers, Songs, and Stories of Healing and Harmony* (Rochester, Vt.: Bear & Company, 2001).

p. 126 *By the summer of 2002, a total of 318 cases. . . . More than a third of the victims died:* Centers for Disease Control and Prevention, "Hantavirus Pulmonary Syndrome—United States: Updated Recommendations for Risk Reduction," *Morbidity and Mortality Weekly Report* 51, no. RR-9 (July 26, 2002): 1–12.

p. 126 *vast majority of victims . . . lived elsewhere in the United States:* Centers for Disease Control and Prevention, Special Pathogens Branch, "Epidemiology of HPS Slideset," 2002; online at http://www.cdc.gov/ncidod/diseases/hanta/hps/noframes/epislides/episl1.htm (last reviewed June 2, 2003).

Chapter 6. A Virus from the Nile

p. 127 *hottest month ever recorded in the city:* Environmental Defense et al., *Global Warming: Early Warning Signs: The Impact of Global Warming in North America* (Cambridge, Mass.: Union of Concerned Scientists, 1999).

p. 128 *the life-threatening phase of his mysterious illness had passed:* J. Robin, "Quotes: On the Evening of August 11," *Newsday*, September 14, 1999.

p. 128 *mosquito bit him one August evening:* J. Robin, "Victim's Final Days: Family Shares Memories with Loved One," *Newsday*, September 8, 1999.

p. 129 *driest stretch in more than a hundred years:* Environmental Defense et al., *Global Warming: Early Warning Signs;* National Climatic Data Center, "Monthly Surface Data for Kennedy

International Airport, April–June," (Asheville, N.C.: National Climatic Data Center, 1999).

p. 129 *heat . . . killed more than a hundred people from the Midwest to the East Coast:* W. K. Stevens, "Across a Parched Land, Signs of Hotter Era," *New York Times*, August 1, 1999.

p. 129 *northern house mosquito . . . often thrives during droughts:* P. R. Epstein and C. Defilippo, "West Nile Virus and Drought," *Global Change and Human Health* 2, no. 2 (2001): 105–107; J. F. Day, "Predicting St. Louis Encephalitis Virus Epidemics: Lessons from Recent, and Not So Recent, Outbreaks," *Annual Review of Entomology* 46 (2001): 111–138.

p. 130 *At dusk they fanned across the borough:* Day, "Predicting St. Louis Encephalitis Virus Epidemics."

p. 131 *his liver and kidneys began to fail:* D. S. Asnis et al., "The West Nile Virus Outbreak of 1999 in New York: The Flushing Hospital Experience," *Clinical Infectious Diseases* 30, no. 3 (2000): 413–418.

p. 131 *first fatality of the mysterious disease:* Robin, "Victim's Final Days."

p. 131 *three more patients with neurological symptoms:* Asnis et al., "West Nile Virus Outbreak of 1999."

p. 131 *official . . . flew to Queens to interview surviving patients:* U.S. General Accounting Office, *West Nile Virus Outbreak: Lessons for Public Health Preparedness* (Washington, D.C.: U.S. General Accounting Office, 2000).

p. 132 *"do everything we can to wipe out the mosquito population":* R. Howell, "Mosquito Coast: Encephalitis Kills 1, Sickens Dozens along East River," *Newsday*, September 4, 1999.

p. 132 *warning residents . . . to remain inside:* Ibid.

p. 132 *developed a paralyzing phobia of flying insects:* D. Morrison, "'Very Paranoid' in Queens, Living with Fear of Virus," *Newsday*, October 2, 1999.

p. 132 *hundreds of crows had begun dying:* U.S. General Accounting Office, *West Nile Virus Outbreak.*

p. 132 *veterinarian . . . treated more than fifty ill crows:* S. Shapiro, "As the Crow Dies: A Bird in Distress: Crows' Deaths Tied to Drought," *Newsday,* September 14, 1999; J. Charos, personal communication, July 2001.

p. 133 *dead crows all over the base:* C. Kilgannon, "At Fort Totten and Elsewhere, Crows Dying Mysteriously," *New York Times,* Sunday, August 22, 1999.

p. 133 *a passerby happened upon four dead pigeons:* P. Dickens, "The Discovery of the West Nile," *Newsday,* September 26, 1999.

p. 133 *a captive cormorant, three Chilean flamingos, a pheasant, and a bald eagle also died:* Ibid.

p. 133 *worst die-off of crows in thirty years:* J. Steinhauer, "Outbreak of Virus in New York Much Broader Than Suspected," *New York Times,* September 28, 1999.

p. 134 *all but ruled out Triple E as the culprit:* T. McNamara, personal communication, August 2001.

p. 134 *all the birds stricken . . . were native to the Western Hemisphere:* Ibid.

p. 134 *challenging the CDC's diagnosis:* U.S. General Accounting Office, *West Nile Virus Outbreak;* Ibid.

p. 135 *Winds gusted to thirty miles per hour:* "History for Eilat, Israel: October 17, 1998," Weather Underground.

p. 135 *flock of 1,200 birds set down at Eilat, in Israel's southern tip:* M. Malkinson et al., "Intercontinental Transmission of West Nile Virus by Migrating White Storks," *Emerging Infectious Diseases* 7, no. 3 (suppl.) (2001), 540.

p. 135 *Israel's government tested a number of wild storks and the geese:* H. Bin, "West Nile Fever in Israel 1999–2000: From Goose to Man," in *International Conference on the West Nile*

. . . .

Virus (White Plains: New York Academy of Sciences, 2001).

p. 137 *agreed to run further tests on the samples:* U.S. General Accounting Office, *West Nile Virus Outbreak.*

p. 138 *this spoke volumes:* Ibid.

p. 138 *the CDC issued a press release:* Centers for Disease Control and Prevention, Office of Communication, "West Nile–like Virus in the United States," September 30, 1999.

p. 138 *matched a sample isolated from the brain of the dead goose in Israel:* R. S. Lanciotti et al., "Origin of the West Nile Virus Responsible for an Outbreak of Encephalitis in the Northeastern United States," *Science* 286, no. 5448 (1999): 2333–2337.

p. 138 *a strain common throughout the Middle East:* M. Giladi et al., "West Nile Encephalitis in Israel, 1999: The New York Connection," *Emerging Infectious Diseases* 7, no. 4 (2001): 659–661.

p. 139 *More than 20 million overseas passengers disembark there annually:* Port Authority of New York–New Jersey, Aviation Department, "Monthly Summary of Airport Activities: August 1999" (New York: John F. Kennedy International Airport, 1999).

p. 140 *In spring and early summer, the songs of warblers . . . fill the wetlands:* M. T. Fowle and P. Kerlinger, *The New York City Audubon Society Guide to Finding Birds in the Metropolitan Area* (Ithaca, N.Y.: Cornell University Press, 2001).

p. 140 *rare sighting of a European widgeon:* Audubon Christmas bird count, Queens, New York, 1998–1999.

p. 140 *they congregate by the thousands before continuing their patient journeys:* G. Waldbauer, *Millions of Monarchs, Bunches of Beetles* (Cambridge, Mass.: Harvard University Press, 2000).

p. 141 *seven of the fifty-nine people hospitalized . . . had died:* D. Nash et al., "The Outbreak of West Nile Virus Infection in the New York City Area in 1999," *New England Journal of Medicine* 344, no. 24 (2001): 1807–1814.

p. 141 *numerous parrots, parakeets, or lovebirds smuggled through New York each year:* R. Lewis, "With Evidence of Lingering Virus in New York Mosquitoes, Investigators Focus on Preventing Possible Outbreak," *Scientist* 14, no. 8 (2000): 1.

p. 141 *passed the infection to a bird that migrated to Queens:* J. H. Rappole, personal communication, November 2001; J. H. Rappole, S. R. Derrickson, and Z. Habalek, "Migratory Birds and Spread of West Nile Virus in the Western Hemisphere," *Emerging Infectious Diseases* 6, no. 4 (2000): 319–328, November 2001.

p. 142 *migrated south along the Atlantic Flyway:* Rappole, Derrickson, and Habalek, "Migratory Birds and Spread of West Nile Virus."

p. 142 *Many of these may have survived an infection:* Centers for Disease Control and Prevention, "West Nile Virus: Vertebrate Ecology," 2002.

p. 142 *sixty-four-year-old woman . . . came down with the disease:* E. De Valle, "A Break in West Nile War as Second Case Found in State," *Miami Herald*, August 10, 2001.

p. 143 *"the mosquitoes are here and will always be here":* S. Wiersma, personal communication.

p. 143 *She recovered and was released from the hospital:* J. Babson, "Closer to Home: West Nile Infects Woman in Keys," *Miami Herald*, August 25, 2001.

p. 143 *virus had infected people in ten eastern states:* L. K. Altman, "Four Are Killed in Big Outbreak of West Nile Virus on Gulf Coast," *New York Times*, August 3, 2002.

p. 143 *potentially alerting people in the disease path:* E. Young, "West Nile Virus Will Sweep across Whole US," on-line at http://www.newscientist.com, 2002; National Aeronautics and Space Administration, "Satellites vs. Mosquitoes: Tracking West Nile Virus in the U.S.," press release no. 02-029, 2002.

p. 143 *identified Louisiana as a potential trouble spot:* D. Rogers, personal communication, July 2002.

· · · · · · · ·

p. 144   *Fifty-eight people fell ill*: Altman, "Four Are Killed in Big Outbreak."

p. 145   *first human case of West Nile virus in California*: Centers for Disease Control and Prevention, "Provisional Surveillance Summary of the West Nile Virus Epidemic: United States, January–November 2002," *Morbidity and Mortality Weekly Report* 51, no. 50 (December 20, 2002): 1129–1133; J. Berck, "National Briefing: Northwest: Washington: West Coast Raven Had West Nile," *New York Times*, October 4, 2002.

p. 145   *"If there's not a mosquito already in California"*: Rogers, personal communication.

p. 146   *Mertyn Malkinson*: M. Malkinson, personal communication, November 2002.

Epilogue: SARS and Beyond

p. 147   *"outlook with regard to microbial threats to health is bleak"*: M. S. Smolinski, M. A. Hamburg, and J. Lederberg, eds., *Microbial Threats to Health: Emergence, Detection, and Response* (Washington, D.C.: National Academies Press, 2003), 245, 247.

p. 148   *"the first severe and easily transmissible new disease to emerge in the 21st century"*: David L. Heymann, testimony at Senate Committee on Health, Education, Labor, and Pensions hearing, "The Severe Acute Respiratory Syndrome (SARS) Threat," April 7, 2003.

p. 148   *Liu Jianlun, a sixty-four-year-old kidney specialist*: K. W. Tsang et al., "A Cluster of Cases of Severe Acute Respiratory Syndrome in Hong Kong," *New England Journal of Medicine*, early on-line publication, April 2003; E. Rosenthal, "From China's Provinces, a Crafty Germ Breaks Out," *New York Times*, April 27, 2003; D. G. McNeil Jr., "Disease's Pioneer Is Mourned as a Victim," *New York Times*, April 8, 2003.

p. 149   *a seventy-eight-year-old woman from near Toronto*: S. M.

Poutanen et al., "Identification of Severe Acute Respiratory Syndrome in Canada," *New England Journal of Medicine,* early on-line publication, April 10, 2003.

p. 149 *Johnny Chen:* E. Nakashima, "Vietnam Took Lead in Containing SARS; Decisiveness, Luck Credited," *Washington Post,* May 5, 2003, A01.

p. 149 *Urbani himself died of the pneumonia a few days later:* McNeil, "Disease's Pioneer Is Mourned"; "China Pneumonia Toll Reaches Thirty-Four," BBC News, March 26, 2003.

p. 150 *It had begun to invade Beijing and, authorities feared, China's hinterlands. It had also arrived in the United States:* E. Hitt, "Early Inklings about SARS," *Scientist,* March 24, 2003.

p. 150 *flew from Hong Kong to Frankfurt and Munich . . . before even suspecting he had contracted a new disease:* K. Bradsher, "Carrier of SARS Made Seven Flights before Treatment," *New York Times,* March 11, 2003.

p. 150 *"a worldwide health threat":* World Health Organization, "WHO Issues a Global Alert about Cases of Atypical Pneumonia: Cases of Severe Respiratory Illness May Spread to Hospital Staff," press release, March 12, 2003.

p. 150 *in China, where dozens of new cases were being reported every day:* World Health Organization, Communicable Disease Surveillance & Response (CSR), "Cumulative Number of Reported Probable Cases of Severe Acute Respiratory Syndrome (SARS): From 1 November 2002 to 30 April 2003," April 30, 2003.

p. 150 *fatality rate, at least in Hong Kong, was closer to 15 percent:* C. A. Donnelly et al., "Epidemiological Determinants of Spread of Causal Agent of Severe Acute Respiratory Syndrome in Hong Kong," *Lancet,* early on-line publication, May 7, 2003.

p. 151 *hence its name, "shipping fever":* J. Storz and X. Lin, *Analysis of Viruses in Shipping Fever of Cattle: Emergence of Respiratory Coronaviruses: United States Animal Health Association 2000 Proceedings* (Richmond, Va.: United States Animal Health

Association, 2000). The main culprit in shipping fever, however, is the bacterium *Pasteurella haemolytica.*

p. 151 *cattle type of coronavirus has even been associated with illness in people:* X. M. Zhang et al., "Biological and Genetic Characterization of a Hemagglutinating Coronavirus Isolated from a Diarrhoeic Child," *Journal of Medical Virology* 44, no. 2 (1994): 152–161.

p. 151 *researchers at the University of Hong Kong:* World Health Organization, Update 64: "Situation in Toronto; Detection of SARS-like Virus in Wild Animals," May 23, 2003; K. Bradsher with L. K. Altman, "Strain of SARS Is Found in 3 Animal Species in Asia," *New York Times,* May 24, 2003; S. Bhattacharya and D. MacKenzie, "Civet Cats Most Likely Source of SARS," New Scientist.com, May 23, 2003. http://www.newscientist.com/news/news.jsp?id=ns99993763; last reviewed June 3, 2003.

p. 151 *"persons who might come into contact with these species or their products . . ."* World Health Organization, Update 64: "Situation in Toronto; Detection of SARS-like Virus in Wild Animals," May 23, 2003.

p. 153 *historically unprecedented:* World Health Organization, "Avian Influenza A (H5N1): Two Further Cases Confirmed in Viet Nam, Overview of the Current Situation, Implications for Food Safety," Update 7, January 24, 2004.

p. 156 *keys to . . . human species:* Robert A. Holt, Ph.D., personal communication, May 2003. Holt is one leader of the team at Canada's Michael Smith Genome Sciences Centre that first sequenced the SARS virus. Holt told me that the analysis does not show evidence of recombination (genetic exchange). Therefore, a random mutation may have been responsible for the virus' ability to infect humans.

p. 157 *antibiotic-resistant disease . . . kills some 20,000 people every year in the United States:* U.S. Congress, Office of Technology Assessment, *Impacts of Antibiotic-Resistant Bacteria,* OTA-H-

629 (Washington, D.C.: U.S. Government Printing Office, September 1995).

p. 158 *we may have done just that with SARS:* Paul W. Ewald, personal communication (March 2004).

Acknowledgments

I wish to thank Charles Halpern, former president of the Nathan Cummings Foundation, for his ideas and enthusiasm that helped bring this book into existence. Henry Ng, the foundation's former executive vice president, offered unfailing enthusiasm and support for the idea during my time as a program officer there. My colleague Elvina Scott was always an enthusiastic voice. I also wish to thank the foundation's trustees; my special thanks go to Adam Cummings.

I wish to thank the many researchers who gave interviews, corresponded with me, or read drafts of chapters. Their comments, suggestions, and corrections were invaluable. Although any remaining mistakes belong to me, my appreciation belongs to them for their efforts in helping translate complex subjects into these six stories.

In particular, Tony McMichael offered valuable comments on the introduction; David Bee, Marcus G. Doherr, Christl A. Donnelly, Peter Stent, and Tom Thorne assisted with the chapter on bovine spongiform encephalopathy; Robert Cooper, Bill Cummins, Eric Delaporte, Sian Evans, Beatrice Hahn, Martine Peeters, and Jean-Christophe Vié helped with the chapter on HIV/AIDS; Fred Angulo, Karen Florini, Cynthia Hawley, Mara Vijups, Patrick Wall, and Linda Ward provided invaluable assistance with the salmonella chapter; Linda and John Beckley, Jeff Dunbar, Richard S. Ostfeld, Sarah E. Randolph, and Ted Stiles helped with the chapter on Lyme disease; James Cheek, H. D. Grissino-Mayer, Ben Muneta, Robert Parmenter, Steve Sternberg, and Ron Voorhees helped with the hantavirus chapter; and John Charos, Jonathan Day, Tracey McNamara, Mertyn Malkinson, Bob Perschel, John H. Rappole, Joshua Robin, and Steve Wiersma lent invaluable assistance with the chapter on West Nile virus.

My many conversations with other scholars who are helping to unravel the connections between ecology and health were invaluable. These include A. Alonso Aguirre, Peter Daszak, Jim Else, Paul R. Epstein, Gretchen Kaufman, Stuart B. Levy, Michael McCally, Ted Mashima, Steve Osofsky, Jonathan A. Patz, Mary C. Pearl, Mark Pokras, and Gary M. Tabor.

I have been blessed to have had caring and gifted mentors throughout my life. Thomas Konsler patiently taught me the art of raising honeybees when I was seven—an experience that galvanized my interest in science; Mrs. Hutton and Joseph Lalley of Gibbons Hall encouraged me at a time in my life when I most needed it; and Ron Bromley, Chuck

Carter, Doc Embler, Pop Hollandsworth, and John L. Tyrer influenced my life enormously when I was a student at the Asheville School. Derek Sarty taught me that the natural sciences could be as artistically inspiring as the expressive arts. Jeremy Dole gave me my first professional writing assignment. George Archibald has encouraged my work for nearly twenty years. Sheila Moffat, David Sherman, Al Sollod, and Chip Stem provided encouragement and support through my years in veterinary school. I am indebted to Michael Lerner and Scott McVay, who have taught me so much. The late Franklin Lowe was an inspiring mentor.

Fritz von Klein has been an everlasting source of friendship and encouragement. A special thanks to Terry Whalen for years of love, encouragement, and support. Many others have encouraged my writing at one time or another, including Renée Askins, Ian Baldwin, Richard Bromfield, Sue Carlson, Lisa Drew, Peter Eddy, Tim Kochems, Pam McCormick, Joni Praded, Mary Suchowiecki, and Lori Tucker. Kim Elliman has given invaluable advice and help over the years. My special thanks to Charles and Esther Krupin for their support.

I have benefited enormously from my association with many generous friends and colleagues over the years, including George Aguilar, Brett G. Anderson, Bob Baldwin, Val Beasley, Beto Bedolf, John Bennett, Ken Brecher, Hooper Brooks, Claudine K. Brown, Clell Bryant, Betsy Campbell, Alexandra Christy, Carlo Colecchia, Andi Colnes, Rachel Cowan, James Cummings, Rick Cummings, Roberta

Cummings, Sonia Cummings, Barbara Knowles Debs, Dianne Dumanoski, Stephen P. Durchlag, Bill Eddy, M. Annette Ensley, Cynthia Evans, Peter Forbes, Joel Getzendanner, Marion Gilliam, Ken Gilmore, Paul Gorman, David Grant, Jim Haba, Leslie Harroun, Robert Hass, Tony Hiss, Dee Hock, Kurt Hoelting, Lloyd Huck, William Hull, Marty Kaplan, Philip C. Kosch, Jane Kretzmann, Andrea Kydd, Rob LeBuhn, Reynold Levy, Nancy Lindsay, Harvey Locke, Lynn Lohr, Kit Luce, Dan Martin, Beatrice Cummings Mayer, Robert N. Mayer, Bill Meadows, Betsy Michel, Elise Miller, Pi'ikea Miller, Suzanne Murray, Pete Myers, Anita Nager, Michael Northrop, Margaret O'Dell, Paul O'Donnell, Sandy Oschelegel, Frank Parker, Robert Perry, Rachel Pohl, Catherine Porter, Will Rogers, Jonathan Rose, Donald Ross, Andrew Rowan, George Saperstein, Jeanne Sedgwick, Kathy Sessions, Brian Sharp, Albert Siu, Ed Skloot, Ted Smith, Ruth Cummings Sorensen, Fred Sowers, Jim Stevens, Ed Thompson, Ken Tomlinson, Mary Evelyn Tucker, Alicia Ushijima, Mark Valentine, Anne Fowler Wallace, Alvin Warren, Debra Weese-Mayer, Kenneth Wilson, and Laurie Lane- Zucker.

I thank my colleagues in the Department of Journalism and Media Studies at the University of South Florida St. Petersburg, who have created an enormously rich and productive atmosphere not only for students to learn the craft of journalism but also for faculty to teach and practice it.

Jonathan Cobb, executive editor of Shearwater Books/ Island Press, immensely improved the manuscript every time

. . . .

he read it. His abiding love of words and his deep grasp of science combine to give him an editorial eye that is at once analytical, intuitive, and imaginative. Pat Harris copyedited this book with amazing clarity, passion, and precision. Chuck Savitt, president of Island Press, encouraged the idea for this book early on. Tom Bruno of Harvard Medical School's Francis A. Countway Library of Medicine provided invaluable support in my quest for articles.

There are those who helped to make this book, and there are those who make my life: Noelle, William, and Anna, thank you for your love and presence.

Index

203

Mark Jerome Walters is trained in journalism and veterinary medicine. A visiting lecturer at Harvard Medical School from 2001–2003, he is now a professor of journalism and media studies at the University of South Florida St. Petersburg. His work has appeared in *Audubon*, *Reader's Digest*, *Natural History* and numerous other publications. A contributing editor of *Orion* magazine, Walters is also the author of the highly acclaimed book *A Shadow and a Song*. He lives on Florida's Gulf Coast.

The Knockout Artist

BY HARRY CREWS

The Gospel Singer
Naked in Garden Hills
This Thing Don't Lead to Heaven
Karate Is a Thing of the Spirit
Car
The Hawk Is Dying
The Gypsy's Curse
A Feast of Snakes
A Childhood
Blood and Grits
Florida Frenzy
All We Need of Hell

The Knockout Artist

Harry Crews

1817

Harper & Row, Publishers, New York

CAMBRIDGE, PHILADELPHIA, SAN FRANCISCO, WASHINGTON
LONDON, MEXICO CITY, SÃO PAULO, SINGAPORE, SYDNEY

FIRST EDITION

Copyeditor: Jean Touroff
Designer: Helene Berinsky

Library of Congress Cataloging-in-Publication Data
Crews, Harry, 1935–
The knockout artist.
I. Title.
PS3553.R46K56 1988   813'.54   87-45848
ISBN 0-06-015893-X

88 89 90 91 92 RRD 10 9 8 7 6 5 4 3 2 1

For Rod and Debbie Elrod,
who made every effort to keep me sane—
and very nearly succeeded—
during the struggle to write this book

This hocus-pocus succeeded:
I buried death in a shroud of glory.

—JEAN-PAUL SARTRE

The Knockout Artist

Other than the city of New Orleans,
nothing in this book is real.
The people do not exist;
the events never happened.

Chapter One

From where he sat on a low stool, the boy—whose name was Eugene Talmadge Biggs, but who was often called Knockout or K.O. or Knocker—had counted the suits hanging in the open closet three times. And each time he counted them he came up with a different number. That did not surprise him. He was not a good counter. It was just something to do until it was time for him to go out and do the only thing he had left. Besides, nothing much surprised him anymore.

He had decided a long time ago that the trick was to try to do the next thing in front of you and not think about it too much. Up to a point, anyway. The trouble was in knowing where the point was, the point beyond which you should not go. Sometimes in the quiet hours of the night when he was lying in bed waiting for sleep, or even walking down the street in the bright light of morning, the thought would come to him that he had long since passed the point beyond which he should not have gone and that his life would never be his own again.

He forced himself to carefully count the suits one more time. And got yet another number. There were probably either 130 or 127 or 133 or 128 suits of clothes hanging in the open closet just there in front of him. And on the floor beneath each suit was a pair of shoes. So however many suits were in the closet, there were that many pairs of shoes also. It occurred to him the first time he saw them that there were not that many suits in all of Bacon County, Georgia, which was where he came from. But he was not in Bacon County now. He was in a house that was as big as a train station on St. Charles Avenue in New Orleans, Louisiana, and things were different here from the way they were where he came from. Christ, were they different. The whole world had changed up on him in New Orleans. Like the houses on St. Charles Avenue. There were few of them he could look at and not be reminded of a train station or else wonder why in God's name anybody would want to live in something so unthinkably huge.

He looked down at the new pair of boxing shoes he was wearing. It was the first time he had ever had them on because they were not his. They had been furnished to him, along with the pair of Everlast boxing trunks he was wearing. They did not fit him very well because he was not wearing a cup, a jockstrap but not a cup, and neither was he wearing the heavy leather belt that would have held the cup. He had no need of a cup in the bouts he fought now. His hands resting on his knees were already taped. He had taped them himself.

He got up from the stool and tested the carpet under his feet.

It was deep, deep enough that a man could sprain an ankle if he was not careful. He shot out a couple of jabs, moved left, then right, always careful of the way he stepped on the carpet.

From the crowd out in the enormous ballroom at the center

2

of the mansion a roar of voices exploded, then faded. The shouts and screams sounded not of celebration but of anger. Eugene told himself that he did not give a good goddam if they were celebrating or going to war. He was standing in the dressing room of a man who owned a hundred-and-some-odd suits of clothes and who was not normal in any way that mattered and whose friends were not normal either so he did not give a good goddam what they were doing or thinking or feeling out there.

The whole wall at the end of the closet was a mirror, and Eugene watched himself bob and weave, move in and out, cover up, and then come out hooking. He could see the bathroom reflected behind him, see the gold handles of the spigots blinking on and off like lights. He did not know if the handles were real gold or not. But it would not have surprised him if they were, and he did not spend any time thinking about it. Now that he was moving, shooting out his left and slipping imaginary punches, he could not help but remember the time when this was all real, even as he heard the roaring voice of the fake fight crowd out in the ballroom of the mansion.

Even though he had not done any road work in a very long time, he was still a notch-solid middleweight, standing nearly six feet two inches tall in his stocking feet. He was startlingly handsome—dark, nose off-center from being broken, and just enough scar tissue in his eyebrows to make him look a little sinister, maybe even dangerous. It was the kind of face that caused people to stop talking in elevators and turned heads in restaurants. He had been aware for a long time of how striking his features were, of how smooth and strong his stride was when he moved along the street or through the lobby of a hotel. He had been asked more than once if he had ever acted or if he was in fact an actor. Sometimes he said he was. Which he knew was not exactly a lie.

He was just about to sit back down on the stool when two men came into the room. One of them was wearing a soft cap pulled low over his eyes and chewing on the stub of a short, cold cigar. The other was wearing a black, tight, sleeveless shirt embossed with a white legend that read: GEORGIE'S IN, WHERE IS YOURS? They were both as thin as fashion models.

"You're Knockout?" said the one in the cap.

"Eugene Talmadge Biggs," he said.

"But sometimes called Knockout or K.O. or sometimes Knocker, right?" said the one in the sleeveless shirt.

Eugene sighed: "Sometimes."

"My name's Georgie. This is Jake." His hands were in constant motion, making little designs on the air. "We're your handlers. My God, Jake, can you imagine making me his handler?" He looked to be in genuine pain. "He's beautiful. I wish you'd look at him, Jake, he's beautiful. They didn't tell me he was beautiful."

"Overlook him, kid," Jake said. "He's obnoxious at times but harmless. I'm your manager. Georgie's your trainer."

It was only then that Eugene realized his manager was a woman. She was an absolute breastless and hipless wonder.

"Jake's a good name for a manager," said Eugene. "Been a lot of great managers named Jake."

"Thanks," Jake said, unsmiling, chewing her dead cigar. "We're ten minutes from cranking up this operation."

Eugene said: "Let's do it and say we did."

"How do you feel, kid?"

"I feel like doing it. The sooner the better. I'm ready to go home right now."

Georgie bent to a leather bag on the floor beside him and came up with a robe.

"What color is that?" Eugene said.

"The royal color," said Georgie.

"It's ugly."

"Purple is the only color for the king."

"I'm not the king."

"You're too modest." He held up the robe so Eugene could

4

see the back of it. Elaborately stitched there in a kind of Germanic script was THE KNOCKOUT ARTIST.

"Wonderful," said Eugene.

"I'm glad you like it," said Georgie. "It was my idea. I think you'll find we run a class act here."

"He was being sarcastic, for Christ's sake," Jake said.

"Oh, hush you," Georgie said. Then to Eugene: "Let me help you on with this."

Eugene shrugged into the robe and Georgie stood massaging his neck. His hands finally slipped to Eugene's back, then to his hips. Eugene stood utterly still, but the muscles in his jaw knotted hard and held tight.

"If you don't stop that, Queen," said Jake, "the king's going to drop you like a bad habit."

"For sweet pity's sake, try not to be so butch," said Georgie, taking his hands away.

"Can we get the fuck on with this?" Eugene said.

"How many knockouts, kid?" said Jake, sliding into a pretty good imitation of a Brooklyn accent. "I, myself, I never seen you work. But I heard. I heard plenty."

"Everybody's heard plenty about the Knocker," said Georgie. "Go on, give Jake the word and watch him faint."

Eugene said: "I got seventy-two."

"A stone miracle," Jake said. "All the fighters in history who had that many you could count on one hand."

Georgie said, "Does that count . . . I mean, does that include before . . . Are you counting the ones when you first started out?"

Eugene, his eyes flat and hard and his face showing nothing, turned to watch him. And then finally: "You asked how many knockouts and I told you. Seventy-two. All right?"

"Take it light, kid," Jake said. "I told you already, obnoxious but harmless. Georgie, the gloves."

Jake took off the soft cap and Eugene saw that she was beautiful. Since she had been standing so close to him and he

5

was so much taller than she was, the cap pulled low over her eyes had hidden her face. But now as she stood turning the cap in her hands, he marveled at her thick hair, red and cut like a boy's, her thin nose with delicately flaring nostrils, her full mouth—when she took the dead cigar out of it—parted to show brilliantly white teeth. It was only then that he realized that she was probably younger than he was.

He deliberately looked away from her and instantly she was gone from his thoughts. If she was caught in a trick of shit, that was her problem, and he was not going to think about it. Dismissing her not only from his head and heart but also from the world was an easy thing. Eugene had had a lot of practice at it. He looked back at her and killed her where she stood. She simply no longer existed, she was gone and whatever problems she might have were gone with her. He stood there shifting on the balls of his feet, his head clear, thinking of nothing, worried about nothing. He knew exactly what was coming, knew that he hated it, but knew also he could and would go out and do it without thinking about it or worrying about it or letting it stay with him very long after it was over. Again, he had a lot of practice at walking through whatever he had to do and not being touched very much by it.

Georgie had taken a pair of red boxing gloves out of the leather bag at his feet and was now lacing them on Eugene's hands.

"Do I pass muster, kid?" Jake said.

"What?"

"You were staring," she said.

"Yeah, I was," he said. "But not anymore."

"You think I'm beautiful, don't you?"

"I don't think about it."

"Striking, then? At the very least, striking," Jake said, giving him a full, open smile, her wet tongue doing something with her upper lip. "Could you get it up for me, Knockout?"

"I'm just doing a job here, lady."

"I'm not a lady."

"No, I guess you're not," he said.

"But I bet you're thinking that if I got fucked right, really fucked right just one time, I would be. Isn't that true?"

"You got to try to believe me when I say I really don't think about it," he said.

"All men think about that, the whole sorry lot of you." Her Brooklyn accent had disappeared. "Lousy beasts."

Georgie said: "I'm supposed to be the obnoxious one, remember? And surely one is quite enough, for God's sake. This young man did not come here to be insulted." He was finishing with the other glove. "You did ask for six-ounce gloves, right, Knockout?"

"Makes it easier," Eugene said.

Jake put her cap back on her head and the cigar in her mouth. "It don't mean nothing," she said, her accent back. "Don't let what I said get to you."

"You couldn't get to me if you wanted to," he said.

"All right, Knocker," said Georgie. "Let's go out and do it. Showtime."

Jake led the way. Eugene Biggs followed, with Georgie behind him. They left the dressing room and went into a master bedroom with a bed larger than Eugene thought most of the rings he had ever fought in had been. It was covered with a leather bedspread under a ceiling that was one solid mirror in a gold-flake frame. Except for the gold-flake frame, the entire room and everything in it was blood-red.

The passageway leading out of the bedroom had Spanish arches and was dim with indirect lighting. At the end of the long passageway was a tremendously bright burst of light and a rising babble of voices. The soundtrack from the motion picture *Rocky* exploded into the hallway and the babble rose to a sustained roar.

Eugene Biggs stepped into a ballroom where perhaps three hundred and fifty people, men and women, milled about or sat on little fight stools. Everybody in the room was dressed as a fighter—Everlast trunks, soft fight shoes, gloves—or as a manager or trainer after the manner of Jake and Georgie. Eugene had the hood up on the robe he was wearing and kept his eyes down as he moved in a little bounce through the crowd. He glanced up once and looked right into the eyes of a woman dressed as a boxer. She was naked from the waist up, her rib cage massive, her breasts the size of golf balls with rigid nipples half an inch long. She was taller than Eugene Biggs and he guessed her to be a light heavyweight. She was being shouted at by a diminutive female manager dressed in a sweat-stained fedora and white shirt with suspenders and a pair of heavy brown trousers that were so wide at the cuff they entirely covered her shoes. An unlit cigarette dangled from her tight, grim little mouth.

Beautiful young black men, every one of them a middleweight or lighter, dressed as sparring partners in sheepskin headgear, their white teeth biting on mouthpieces, moved through the crowd with silver trays full of drinks in long-stemmed glasses and freshly rolled joints and little clear, cut crystal bowls of pills and capsules of every imaginable color and coke laid out in lines. Tables at either end of the room were piled with food, some iced, some steaming.

Eugene Biggs didn't see the ring until he was actually about to climb up into it. The ring posts were covered in velvet and the ropes were velour. The ring itself was hardly bigger than a pool table. Jake put her foot on the middle rope and pulled up on the top rope. Eugene stepped through. The theme from *Rocky* thundered on. The men and women dressed as fighters who had been sitting on little stools got up and pressed forward to the ring along with the managers and trainers. Even the sparring partners put down their silver trays and pumped their taped fists above their heads.

His robe was removed and Eugene danced across the ring,

giving two lightning-quick Ali shuffles, showing the great foot speed he had always had. Back in the corner, he glanced down at the water bottle, a huge Perrier in a champagne bucket. The spit bucket sitting beside it was a tall silver wine chalice.

He leaned near Jake and pointed to the Perrier and chalice. "That shit ain't necessary."

It was Georgie who answered. "When Oyster Boy gives a theme party, he insists on the authentic, the pure."

"Oyster Boy?" said Eugene.

"You didn't meet him yet?" said Jake.

"No."

"The matchmaker," said Jake. "The promoter. This is his place, his arena."

"He's called Oyster Boy," said Georgie, "because . . ."

"Save it," said Eugene. "I don't want to know."

"You'll want to know this," said Georgie, grinning for all he was worth. "It's because he . . ."

Eugene spun and caught Georgie behind the neck with his gloved hand. Their noses were almost touching. "I said save it, asshole."

"Oh my God," said Georgie, his grinning mouth going slack, "if you hit me, I'll come."

A young man wearing a tuxedo came twirling and cat-dancing into the ring. His blond hair reminded Eugene Biggs of a movie he had once seen with General George Armstrong Custer in it. A microphone descended on a wire from the ceiling and the young man seized it. He pressed it against his pursed, beautiful lips.

"Ladies and gentlemen!" A mild roar came out of the audience. "Dirt track specialists!" The applause and noise of voices instantly went a notch higher. "Fags and fagettes, clits and slits and meatpole mountaineers!" Now, even through the amplified speakers, he could hardly be heard for the applause and shouts of approval.

9

"I'm at a fucking AIDS convention," said Eugene.

Georgie, who had been massaging his shoulders, said: "That's cruel and unfair."

"Who told you it wasn't going to be cruel?" Eugene said. He rolled his shoulders away from Georgie's soft, probing fingers. "And fair? I never got any mileage out of fair. Ain't nobody in this room ever got any mileage out of fair. Did, they wouldn't be here." He smiled at Georgie, who looked as though he was at the edge of tears. "But don't spend any time thinking about it, Georgie."

"This is the main event of the evening," said the young man with the long yellow hair into the microphone. "Ten rounds of boxing in the middleweight division. In this corner"—he pointed to Eugene Biggs—"weighing in this afternoon at one hundred fifty-nine and one-half pounds . . . with a record of seventy-two and 0 . . . all by knockout . . ."

The audience had taken up a steady chant of Knockout, Knockout, Knockout, Knockout, Knockout.

". . . a big hand, please, for New Orleans' favorite fighter, Eugene 'The Knockout Artist' Biggs!"

Eugene danced out of his corner into the steady, pulsing chant of Knockout, Knockout.

"And in the blue corner . . ."

Eugene Biggs backpedaled into the blue corner and the audience fell utterly silent, so silent that the only sound in the room was the breathing of the announcer into the microphone. ". . . weighing in at precisely the same weight this afternoon at one hundred and fifty-nine and one-half pounds . . . a fighter much better than his record because he has always faced fighters the caliber of The Knockout Artist . . . a record of 0 and seventy-two . . . let's give a warm Oyster Boy welcome to Phil 'Fight and Die' Phillips."

Silence still, the only sound that of the announcer's breathing into the microphone. Eugene looked out over the grim faces of the audience. They looked like they wanted to lynch him or gut him and eat his heart or set him on fire.

"The referee for the fight will be our own Russell Muscle." Eugene Biggs had not seen Russell Muscle climbing into the ring but, looking at him now standing in a neutral corner, he didn't see how he had missed him. An iron freak, thought Eugene Biggs, an honest-to-God iron freak.

Russell Muscle was at least six four, probably two hundred and sixty pounds with what looked to be a twenty-eight-inch waist, twenty-two-inch arms, and a fifty-two-inch chest. His thighs, which had to be bigger than his waist, seemed about to split the white linen trousers he was wearing. Eugene Biggs had worked out in many gyms where body builders trained, worked out around them enough to be a good judge of their weight and how they would tape out. And also to hold them in contempt, thinking of them, as he did, as mirror athletes.

The iron freak called him out to the center of the ring, turned him to face the audience, and then stood behind him speaking to the back of his head. The audience was still grim and silent.

"You had your instructions in the dressing room. Good luck to both of you. Shake hands and come out at the bell."

Eugene slapped one of his gloves on top of the other, shaking hands with himself, and danced back into the corner where Jake and Georgie waited, standing now outside the ropes on the ring apron.

"The Oyster Boy ought to be in show business," said Eugene Biggs, speaking more to himself than to either Jake or Georgie.

"Oh, he is," said Georgie. "He's very big in show biz."

Jake reached out and pressed a mouthpiece against his lips.

"That ain't necessary," said Eugene Biggs.

Jake said: "Take it. Oyster Boy insists on the authentic, on the real thing."

"You already told me, but I'll only have to lose it later."

"Lose it then, but take it before the bell."

Eugene opened his mouth and took it and when the bell sounded he flashed into the center of the ring, moving left and then right, putting himself in a corner and then spinning out. Spontaneous and synchronized, a chant came up as a single voice: *Knockout, Knockout, Knockout, Knockout.*

He showed more swift lateral movement and then went into a corner again. This time when he spun out, he went flat-footed in the center of the ring and set himself to punch. He blew the mouthpiece entirely out of the ring and at the same time let his jaw go slack. He caught himself on the point of the chin with a vicious right cross. The lights dimmed, his knees buckled, and the chant *Knockout, Knockout* moved far away as though he were hearing it through a wall from another room. Even as he was going down and the lights were no longer there, a voice, his own and hollow as an echo, spoke out of the darkness: "Well, it's over one more time."

He came to with an ammonia capsule broken under his nose and a young man shining a penlight in one of his eyes.

"I'm the ring doctor," said the young man, holding up two fingers. "How many do you see?"

"Two. Get away from me."

"You knocked yourself cold as an Eskimo's balls, you." He had a Cajun accent.

Eugene Biggs sat up and shook his head. "I know what I did. Get away from me and let me get up."

"Actually, I'm not a doctor. Yet." He giggled, and Eugene saw for the first time that his cheeks were rouged and he was wearing lipstick. "I'm a med student at Tulane."

Eugene said: "Good for you." He struggled to his feet as Jake and Georgie moved in to support him. "I can make it by myself," he said.

He never acknowledged the wild cheering while he was in the ring or climbing through the ropes or even as he made his way out of the room, Jake in front of him and Georgie following, through a narrow aisle the audience opened for him. Hands reached out to slap his shoulder or back, and more than a few his ass, but he kept the hood of the robe up and his eyes down.

At the end of the passageway, he let Jake lead him to the bed, where he sat on the leather spread.

"You feel all right?" said Jake.

He looked at her. She wasn't a bad sort, he thought. There was genuine concern in her voice. And genuine concern was rare enough in his life, had been ever since he left home.

"Yeah," he said. "I'm all right."

And that was the truth. There had been no pain. There never was. There was only a feeling of tremendous pressure at the base of his skull as his jaw unhinged from the punch. Then there was nothing. Afterward there was sometimes a headache and slight nausea, as there was now, but nothing he couldn't handle.

"Can I get you anything?" said Georgie.

"No," he said. And then: "Well, yeah. My money."

But even before he had finished speaking an enormously fat young man wearing an Adidas warm-up suit came into the room. He had chin after chin rolling down his chest toward a ballooning stomach that ended in a flap of fat even the warm-up suit could not hide. It hung across his thighs like an apron. His eyes were no more than slits in his swollen face. He had a leash in his right hand, the end of which was attached to a leather collar decorated with steel studs and fastened about the neck of an extremely thin man whose head was entirely bald, showing not a single wispy strand or a trace of stubble. He was dressed as a boxer and was as tall as Eugene but could not have weighed more than ninety-five pounds. Every bone in his body was insistent under his skin, skin that was diaphanous and desiccated. Eugene couldn't take his eyes off him. He was the most unhealthy-looking human being he had ever seen. As he watched, the man reached up and scratched his

13

chest. A little shower of dead skin fell to the thick red carpet.

"Knockout," said Georgie, "may I have the honor of introducing you to Oyster Boy."

Eugene looked at the boy, whose fat made his age hard to guess, but Eugene thought he couldn't be more than eighteen, maybe even less. He acknowledged him by simply repeating his name: "Oyster Boy." He did not offer his hand.

"I'm Oyster Boy," said the emaciated man in the steel-studded collar at the end of the leash. His voice was as dry and scaly as his skin.

Eugene Biggs closed his eyes and raised his hand to massage the base of his skull, where a dull ache throbbed like a pulse. He wanted very badly to get his clothes and get out into the street. The room had begun to feel tight and airless even though it was chilled to the point where he thought you could keep fresh meat in it.

"Have you got a headache?" said Oyster Boy.

Eugene opened his eyes and shook his head.

The fat boy said: "We've got Tuinal, Nembutal, Seconal, even dilaudid, if you want to really go down."

"All I want is my money," said Eugene Biggs.

"Pay him, Purvis," said Oyster Boy.

The fat boy reached into the pocket of his warm-up suit and pulled out a roll of bills. He snapped off ten one-hundred-dollar bills and held them out to Eugene.

"Put it on the bed," Eugene said.

Purvis put the money on the bed. Eugene picked it up and stuffed it into the waistband of the Everlast trunks.

"Wasn't he sensational out there," said Georgie.

Oyster Boy didn't look at Georgie. "I thought it would last longer."

"The guy I made the deal with . . ." said Eugene.

"Joey Q," said Oyster Boy. "You made the deal with Joey Q. He's my agent for certain . . . certain things."

"Yeah, well, he didn't say anything about time and I didn't

either. He knew what I was going to do, and if he's your agent, you should have known, too."

Oyster Boy's lipless mouth, thin as a razor cut, went thinner still in a smile. "Young man, I was not in any way criticizing your performance. It was only an observation. Now that the fight is over, won't you come out and have something to eat with us?"

Purvis put one hand on his enormous belly and said: "We've got food to the end of the world out there. Panéed veal with czarina sauce, roasted quail stuffed with crawfish dressing, corn maque choux, stuffed mirliton with shrimp and crab sauce." The names of the dishes were tumbling more and more rapidly from his mouth, which was getting wetter with every word. "Rabbit tenderloin with mustard sauce, tasso and oysters in cream on pasta that will break your heart, fried oysters bayou teche . . ."

"Purvis!" said Eugene. "Put a lid on it."

Purvis stopped, startled, his wet mouth open. He had fallen into a kind of trance over the catalogue of food. "Lid?" he said. "Which lid?"

"Lose the menu," said Eugene. "I ain't eating."

"You will disappoint my guests," said Oyster Boy. "They were looking forward to . . . to meeting you."

"Not the deal," said Eugene Biggs. "The deal's done."

Oyster Boy said: "You mustn't take that tone, Knocker. You quite made the party. It was . . . unique."

"How did you know about me? How to find me?" Eugene Biggs was here because a man, well-spoken and straightforward, had knocked on his door a week ago and given him five hundred dollars up front.

"Very little happens in New Orleans or, for that matter, Louisiana that I'm not aware of," Oyster Boy said. "You're better known than perhaps you realize. Some of my guests claim to have actually seen you . . . seen you fight before tonight."

"I could give a rat's ass," Eugene said. "Just wondering."

"There's that tone again."

"You do seem to be copping an attitude," said Jake, who had been lounging against the far wall, examining her nails, occasionally biting one with her sharp, perfect teeth.

"I ain't copping nothing but my clothes and I'm out of here. It's finished."

Oyster Boy looked to Purvis. "I think it's time."

Purvis turned his head toward the door opening into the passageway and screamed: "Oysterrrs!"

As if by magic, two slender young black men wearing sheepskin headgear, their teeth clamped on mouthpieces, popped through the door carrying wide silver trays on their shoulders. The silver trays were mounded with huge, wet oysters, some as thick as a man's wrist.

Oyster Boy's mouth showed the razor grin again. "They don't call me Oyster Boy for nothing."

Eugene Biggs was off the bed and had his clothes out of the dressing room before the sparring partners could set the trays down.

"Wait, Knockout," said Oyster Boy. "We're just going to have a little fun. You'll hurt my feelings if you leave." His voice was full of pleading, maybe even of grief.

"My deal didn't include oysters," said Eugene, pulling on his trousers over his boxing trunks.

"My God, a virgin," said Jake.

"Yeah, lady," said Eugene, slipping into his motorcycle boots, "just you and me, both virgins."

"You'd be surprised what you can do with an oyster," said Purvis.

"No, I wouldn't," Eugene said, buttoning his shirt as he left the room.

16

Chapter Two

Three months before his eighteenth birthday Eugene Biggs had left high school, and left his daddy's small farm not far from the Okefenokee Swamp in South Georgia, left because it was expected of him. Graduating from high school was not a consideration; nobody in his family had ever graduated from high school, and besides that he had never known anybody who had made a dime from book learning, and he did not care for it anyway, had never been good at it. Nobody had ever told him he was expected to leave but he had always known it. He was grown and now he had to make his own living, graduating or not graduating had nothing to do with it. His daddy's land was nothing but a thin layer of leached-out topsoil over a bed of solid clay, and there was not enough of that in cultivation for him to stay at home.

His older brother had managed to get a job cutting right-of-way for a power company, and had moved into a rented room in Alma, the county seat of Bacon County. But that had been entirely luck. There simply were no jobs to be had. Maybe work here and there for a day or two, or even a week, but nothing steady.

So he did what the people in South Georgia had been doing since before the Great Depression, what his own daddy had done until he could save enough money to buy a piece of dirt with a heavy mortgage, a mortgage he had carried ever since. He got on the Greyhound bus and went to Jacksonville, Florida.

He took a rented room on Eighth, just off Main, and in a week he was working at the Merita bakery on the loading docks. He worked there two months before getting a job in a pulp mill on Talleyrand Avenue which paid a little better. From there he went to work for a roofing company spreading hot tar with a mop, and from there to a construction company mixing cement and pushing it in a wheelbarrow to bricklayers.

He had no skills at all, so when he searched for work, for a job that paid a little more than the one he had, his body was the only thing he had to sell. But he was young and lean and strong and anxious to work.

"Just put me on one day," he would say, "and if I don't suit you, you don't owe me nothing and I won't come back anymore. It ain't nobody on this job can work me down. I can put anybody here in the shade. Put me on and see if I caint."

So he had not had any trouble getting work, but then one night his daddy's brother, Carter Biggs, had come to see him in his room.

"Everything all right, son?"

"Yes sir, I been getting along fine down here."

"You still going from pillar to post all over town, one job to the next?"

His uncle Carter had come to Jacksonville the same time his

daddy had and, after a year of trying, got a job at the Jefferson Shipyard Company on Bay Street that runs parallel to the St. Johns River. He had learned to weld and had been in the shipyard ever since. He did not approve of his nephew bouncing from job to job.

"I'm mixing mud right now," Eugene said. "Ain't bad work, Uncle Carter."

"Ain't bad work now while you a young buck. Try it when you forty-five."

"I won't be doing it when I'm forty-five."

"What you plan to be doing?"

"I don't know."

"Damn right you don't know. And if you don't learn how to do something you gone end up taking handouts from the guvment. I told you when you come I was gone hep you if I could. Don't go to that mud job tomorrow. Go to the yard and see the foreman in the sheet metal shop. He'll put you on as a metalworker's hepper."

"I don't know a thing in the world, Uncle Carter, about sheet metal work."

"You don't have to know nothing to be a helper. Helpers just tote things, do what the journeyman tells you. This is a real chance for you, and you better get on and damn well stay on. The foreman's a old boy who comes from up there by Ten Mile Creek and I been talking to him ever since you daddy told me you was coming. You got the job if you at the yard in the morning at seven-thirty."

Eugene was at the yard before seven and when he saw the foreman, he got the job. The foreman took him over to Budd Jenkins, a short, thick man with a cap of tightly curling gray hair, a bridgeless nose, and a meaty face filled with scar tissue.

He hardly looked at Eugene. "Grab the toolbox, boy."

Eugene took up a metal box with a leather handle and rushed to catch up with Budd Jenkins who, despite his age, moved over the broken ground littered with metal shavings, pieces of angle iron, and parts of steel plating, with a smooth stride that was hard to keep up with.

They were supposed to replace a rusted flour bin in the galley of a ship in dry dock, but as soon as they had climbed to the galley, Budd turned to him and said: "All right, Biggs, your first job. Try not to fuck it up. Keep a eye out for me, I got to take me a little nap. I don't feel so good. You see somebody coming or the ship catches on fire—I mean really on fire—wake me up."

Budd lay down on a stainless steel table and in less than a minute he was snoring. The air horn for the lunch break woke him up. He got off the table, yawned, stretched, and then walked over and opened his toolbox. There was a lunch bucket in it. The only tools were a hammer and a screwdriver.

He looked over at Eugene, who had been watching the river for the last four hours and thinking this was the easiest money he had ever made and also thinking that it was dishonest, even maybe against the law for all he knew.

"Bring your lunch?" Budd asked.

"No sir."

"It don't matter. I'll give you something out of my bucket." He handed Eugene a sandwich. "Ever box, kid?"

"What?"

"Box. Did you ever box? You know, fight in the ring?"

"No sir."

"You know anything much about it?"

"No sir, I don't."

Budd grinned: "I do."

And for the rest of the day, that's what they talked about, boxing. They never went near the flour bin, never even mentioned it. Budd Jenkins told him all about managers, the good ones, the bad ones, the ones that brought their fighters along in the right way and stood by them, and the ones that overmatched their fighters for a quick buck. And fighters. God, did he talk about fighters. On his feet, bobbing and weaving, he talked about what they had, what they didn't have, who they were.

"Some got heart, some got talent, some got both. Take Rocky Marciano. Everybody that ever fought him, beat him. But he retired undefeated, right?" Eugene didn't answer. He didn't say anything much all afternoon, except for an occasional grunt or yes or no. But that seemed enough for Budd. "Marciano would take five punches to land one. But Jesus, what a hitter. Archie Moore was the only guy to ever take him off his feet. He hit Marciano with everything but the ring post. But he got off the canvas and whipped the shit out of Moore. All on heart. There was no quit to him.

"And Willie Pastrano, light-heavyweight champion of the world. Couldn't hit for shit. All points, no punch. But a win's a win, that's what they say.

"Then there's guys like Scott LaDoux. He'd maul you, brawl you, butt you till you bled, step on your feet, and thumb you blind. But he only wanted to win. And he was only doing what the referee let him do. It's up to the referee to control the fight. But he would have used his teeth if he could've got away with it. I never saw a referee that could control LaDoux. Last I heard he was refereeing wrestling matches. Jesus, wrestling matches, after being in with some of the best fighters in the world. Well, what the hell, we all gotta do something."

He ran through a catalogue of fighters that most people had never heard of but who were world-beaters, despite never beating the world, never getting the recognition and the money they deserved.

"Take Cleveland Williams. You remember the Ed Sullivan show on television, Biggs?"

"No sir," said Eugene.

"Don't matter. I keep forgetting how goddam old I am. Ed Sullivan had Sonny Liston on when he was champion. Ed says to Sonny, he says, 'Who hit you the hardest since you been

21

fighting?' And Sonny says, 'That be Cleveland Williams.'
'Who?' says Ed. See, Ed Sullivan never heard of Cleveland
Williams, and nobody remembers him today. Great fighter
with lousy luck and lousy managers. But Sonny just says again,
'Cleveland Williams.' 'What did he hit you with?' Ed says.
Sonny Liston says, 'He hit me with a right hand, he hit me with
a lef hand. An I think he kicked me two or three times, too.'
Well, Ed laughed and the audience laughed, but Sonny didn't
laugh. His face looked like he could still feel those punches,
which he probably could."

And he talked about fighters who did everything wrong or
nearly everything, and still won, even became legends.

"Muhammad Ali comes to mind. Half the time he had his
hands at his hips, and he always carried his left too low. By all
rights, somebody should have taken his head off, but nobody
did. Ken Norton and Smoking Joe Frazier come close to doing
it, but the point is they didn't. And Ali's habit of leaning
straight back from a punch. You never lean straight back from
a punch. A good fighter'll kill a guy that leans like that. But
nobody killed Ali. Why is that? I'll tell you why is that. With
Ali's speed, you can carry your hands anywhere, in your fucking pockets if you want to, and you got his balance it don't
matter which way you lean or how much."

And then Budd started talking about fighters coming from
anywhere and everywhere, with good backgrounds and bad.

"If you're carrying the right shit in your fists or heart or both,
you can come from anywhere if you lucky enough to have the
right people see you and bring you along. Think about a wel-
terweight like Mark Breland and a heavyweight like Mike
Tyson."

Eugene could not think about either one because he had
never heard of them. But he sat, quietly nodding his head, on
the edge of the flour bin they were supposed to be replacing.

"Sure, Breland comes from Bedford-Stuyvesant. A goddam
war zone."

Eugene had no idea where Bedford-Stuyvesant was, but Budd was really bobbing and weaving now, and hooking with his left hand, a little froth of spit growing at the corners of his mouth, and Eugene was not about to ask where Bedford-Stuyvesant was.

"But Breland had a solid mother behind him who kept a solid home. He got off the street and into a gym when he was young where the right people, good fight people, saw him and he went all the way to a gold medal in the Olympics. He was protected and managed right all the way, and as a pro he's undefeated, the class of his division. Of course it don't hurt a thing that he's a head taller than anybody he fights and always got a five-, seven-inch reach advantage.

"Mike Tyson? The best young heavyweight in the world. Where was he when Breland was in the gym? In the goddam reform school, that's where. Just a kid, and he's already locked up. Apprentice fucking criminal is what he was before he was old enough to jack off. But who happens to see him? Cus D'Amato, who was Floyd Patterson's manager, took Floyd all the way to the championship of the world. What does Cus do? Nothing except take Tyson home with him, raise him, train him, treat him like a son, even adopted the fucking kid, and made a fighter out of him that's as savage as anybody who ever pulled on a jock. Undefeated and likely to stay that way, Tyson is. And it don't matter a damn that he ain't built like a heavyweight. That's what Smoking Joe Frazier said, not built like a heavyweight. But how did Smoking Joe's son, Marvis, who *is* built like a heavyweight, do against Mike Tyson? Do you know? Do you even know, son?"

"No sir," said Eugene, "I don't." But he wanted to know. He had never met anybody like Budd Jenkins, never seen this kind of passion and enthusiasm in his life. It was this kind of love for something, anything, that he had always wanted, but he had not known it until that afternoon in the galley of a dry-docked ship, the first ship he had ever been on.

"Mike Tyson knocked Marvis Frazier out in thirty seconds of the first round," Budd said with satisfaction. "At twenty-one Tyson owns part of the heavyweight championship of the world!"

"The first round you say? Well, I'll be go to hell," said Eugene because he did not know what else to say.

"It never pays to take a guy light. Or to give him the edge. Sometimes you give him the edge without even knowing it. Take what happened when Emile Griffith met Benny 'Kid' Paret in Madison Square Garden. If you think you can get to a guy by bad-mouthing him and make him mad and rattle him so he'll fight a sloppy fight, then bad-mouth him. Right? But never put the mouth on a guy unless you know enough about him to know how he's going to react. Otherwise, you might give him the edge. When Emile Griffith and 'Kid' Paret met at the weigh-in, the kid starts lipping off about how Emile is a fag, a sucker of dicks, because see, Emile Griffith had this high voice and on top of it he designed hats, women's hats. Can you believe it? Anyway, he did. So the kid really gets in his face about pitching and catching and all that shit. Thing was, it only made Griffith mad as murder, and it happens that he fights better mad than any other way. So when Emile got the Kid into the ring he killed him." Budd beamed upon Eugene and waited for an answer.

"Killed him?"

"Beat him to death. See, Griffith got him in a corner in the twelfth round and the Kid was out on his feet but he couldn't fall because Griffith had him hung on the top turnbuckle and kept him up with punches, just with the power of his punches kept him nailed to the turnbuckle while he killed him." Budd hustled his balls and looked out across the river. "I guess Kid Paret will think a long time before he bad-mouths somebody else."

"I guess," said Eugene.

And that was the way the afternoon went right up to the

24

time the whistle blew for the shift to change. It was not until the next day that Eugene learned that Budd had been a fighter, ranked seventh in the world before a broken right hand healed badly and he had to quit with a record of thirty and three. Budd spent all day on his feet going through all his fights from the first bell to the last, while Eugene sat on the flour bin and listened.

"I was a contender, son. Seven in the world, a contender."

And it wasn't until the third day that Budd told him he spent his free time at a fight gymnasium on Forsyth Street looking for a young fighter he could manage, one with heart or talent or both, one he could take all the way to the top.

"Somewhere right now," said Budd Jenkins, "there's a kid walking around who's a champion. He don't know it, but he's a champion. I'll know it, though. When I see'm I'll know it. I had a couple of boys I thought was the goods. Somehow I was wrong. But you can't quit because you're wrong. I'll find him yet. He's out there, and I'll find him."

Then on Friday when they were drawing their pay, Budd turned to him and said, "You can come down to the gym if you want to. You wanta do that?"

"Damn right I do," said Eugene.

Since he had found out about the gym, he had wanted to see it, but he had been afraid to ask Budd to let him come. Budd was strange, sometimes very strange. When he was deep into talking about fighters and fights, his eyes often went a little crazy. Eugene had thought a lot about the look in his eyes and the only word he could think of to describe it was crazy. But he wanted to go anyway, to see the gym, to see the men who had made themselves special in the world by living with punishment and pain, giving and receiving it.

That moment he walked into the gym with Budd in the late afternoon would live forever in Eugene Biggs because his life changed direction and he was never the same again. The gym was bigger than most barns he had ever seen and had no

25

ventilation except two huge ceiling fans. It was close and hot, hotter than the stainless steel galley on the dry-docked ship, and motes of dust floated in the rays of sun that fell through the high closed windows on the west side of the building.

The whole place smelled of sweat and alcohol and liniment. Fighters moved through the thick, heavy air, dancing through whirling skip ropes, shadowboxing against the bare unpainted walls, doing sit-ups and leg lifts, hitting speed bags that were only a blur in front of their fists that were only blurs, too, or moving in and out on heavy bags delivering punches that landed with an unbelievable sound, a sound that made Eugene think of broken ribs. He told Budd that he did not see how anybody could stand to be hit like that.

"Conditioning," Budd said. "But you don't even feel punches in the fight. The next day, yes. Not in the fight though. You don't feel anything in the fight except maybe the first couple of punches. But, son, the real point is to not get hit. Stick and move, go side to side, develop a left jab that'll keep a guy off you. Too many goddam fighters train to take a punch. Fuck taking. In this business, just like the Good Book says, it is better to give than to receive."

Two fighters were sparring in the big ring at the center of the gym. Eugene had never seen a boxing match, but the fighters seemed clumsy to him.

"They are clumsy," said Budd. "You don't walk out of the woods a fighter, you have to be schooled. But nobody's going to school those two guys. I've seen'm in here the last six months and what you see right there is what I saw the first time I ever laid eyes on them."

They walked over to the ring and leaned on the apron, where a fat man holding a stopwatch and sweating more than the fighters yelled instructions, and encouragement, and insults.

"How they coming, Morie?" said Budd.

"I don't know," Morie said in a disgusted voice. "I don't fucking know."

Budd leaned toward Eugene and said in a low voice: "Bad fighters with a bad manager. Perfect. All undercard shit. Way under."

Later, there were better fighters sparring, and Budd explained what they were doing right and what they were doing wrong. He talked about stances, about foot and hand speed, about hooks and crosses and doubling up, about kidney punches that would make your opponent piss milk for a week, and about how to make your punches come all the way from the bottom of your right foot so that your entire body was behind it. He talked right on through the afternoon and evening until the gym closed at eleven o'clock.

The following Monday, Budd did not ask him if he wanted to go back to the gym with him, but Eugene went anyway, and again they walked directly to the sparring ring and again Budd talked steadily until closing, sometimes turning from the ring to point out fighters doing twists and sit-ups and crunches, and arches off their backs to strengthen their necks.

Back in his room, Eugene started doing the exercises he had seen. He did not know quite why he was doing them but somewhere in him there was the vague notion of someday taping his hands and pulling on a pair of gloves, so it did not surprise him at all when Budd asked him if he wanted to get in the ring and try his hand at sparring. In fact it seemed the next natural and inevitable step, as though they had planned it. He sparred with a club fighter who was fifteen and eleven, and when he came out of the ring Budd asked him if he wanted to be a fighter. The answer was easy.

"That's what I want to be," he said. "It's crazy but that's the way it is. It's crazy but it seems like that's what I always wanted to be and just didn't know it."

"You're right about being crazy. You ain't worth a shit

. . . now. But I know what I know and you could be. I wouldn't waste my time and money on you for any other reason."

The next day after work, Eugene started training, and after a month he could double, even triple with his jab, and he could hook with either hand and became especially good at dropping his right over his opponent's left jab. But mostly what he had was speed, the ability to move, and he also had the ability to see and anticipate.

He had his first fight at the Beaver Street Arena in Jacksonville with an over-the-hill ham-and-egger who gave his age as thirty-six. Eugene won on points.

"A win's a win," beamed Budd Jenkins back in the dressing room. "We'll take it on points or any way we can get it. You got no punch, kid, but it'll come."

But it never did. After winning two more fights at the Beaver Street Arena, Eugene won in Atlanta and in Atlantic City and in Detroit. All on points. After six fights, he stepped up in class and won twice in Jersey City. On points. Budd worked with him on his power, doing everything he could think of, including punching underwater in a pool as Marciano had done to develop his power, but none of it worked.

"So you're not a banger," said Budd. "Don't let it get to you. It's plenty going for you anyway. You're a smart fighter with twice the speed any fighter deserves. And you don't cut and you got a good chin."

He did have a good chin, a chin that had kept him in the fight more than once. As he fought better and better fighters, he had taken some tremendous shots and none of them had hurt him. Not only had he never been down, he had never even been dazed. Most of the time he could not be hit because he could not be caught, but those times he had been tagged didn't bother him.

In the early morning Eugene did six miles of road work, and in the evening he worked out at the gym. At the shipyard, he spent his time napping or watching Budd do what little work had to be done, because Budd insisted that he rest.

He moved in with Budd and ate better than he had ever eaten in his life, a lot of steak and potatoes and steamed vegetables and dried peas and beans, all of it prepared by Budd, and he drank enormous quantities of orange juice. Eugene had not kept a record of it, but with the food, the airline tickets, the equipment and gym workouts, he knew they had not made a dime, but worse than that, he thought he was putting Budd pretty deep into the hole.

"I'm breaking you the way I'm eating," Eugene said, cutting into a sirloin at the dinner table one night, "and besides that, I think . . ."

"Don't think," said Budd, taking a bite out of a raw onion and digging into a can of tuna. "That's what you got me for. I think, you fight. That's the agreement, am I wrong?"

"That's the agreement," said Eugene, but it made him feel terrible all the same. "But what I was going to say is I think we could save money if I didn't . . . if you didn't . . . what I mean is I don't need any . . ."

"You just think you don't need any," said Budd, good-naturedly. "And don't stutter when you talk about pussy, son."

Eugene had not been able to say it, but that in fact was what he was talking about. After every win, Budd had insisted on buying him a whore. He had told him that from the beginning.

"Every man needs pussy," Budd said. "But I'm old-fashion about fighters and fucking. The only time you fuck is when you fight and win. No other time. Ever. That way, you get what you need, and I've got a lock on it. I won't have you willy-nilly dipping your wick here and there. Fucking's got to be a discipline just like everything else."

Thus far in his life, he had had thirteen women, all whores bought by Budd. The encounters had been brutally anonymous and mercifully short. Other than several equally brief encounters with a variety of farm animals, everything he knew about sex he had learned from Budd's whores.

"Some managers make monks out of their fighters," Budd said. "Does damage, too. Be thankful I got understanding."

"I am," said Eugene. "I am thankful."

"All right," said Budd. "You can be thankful. Just don't think. Thinking is the worst thing a fighter can do."

Budd, who seemed to know not only all the fighters and managers in the country but also all the promoters, was on the telephone constantly, talking about the boy named Eugene Biggs he had discovered.

"We got to showcase you, kid," Budd kept saying. "We got to get your name out there somehow. You paid your dues. Don't worry, it'll happen."

And it did. The phone rang one evening while they were eating dinner. Budd answered it and then didn't say a word for a very long time.

Finally he said, "He's ready, we'll do it." Silence again, Budd listening. "He's always in shape. He'll give you a good show. How long I know you? Fifteen, sixteen years? I ever tell you wrong? You damn right I don't. It's done then . . . and, Joey, thanks. Take care."

When he turned from the phone and sat at the table again, he looked grim. "Bad news?" said Eugene.

"It may be," Budd said, "or it may be the best news we'll ever get. It depends on how you do. A boy fucked up his hand in his last workout before a match and had to pull out." He smiled, but even the smile looked grim. "They want you. Think you can go Saturday night?"

That was only two days away, but Eugene knew he was ready and felt the surge of excitement and adrenaline that always came when he knew he was about to fight.

"You know I can go. Where is it?"

"How do you feel about New York? On the undercard at Madison Square Garden."

How he felt was stunned and sick. He could not even answer.

"Don't get freaky on me. It's just a fucking fight. And a fight's a fight. It don't matter if it's in a barn or in the Garden on national TV."

"Lord help me, is this on TV?"

"The boy you're going against is a real comer. A lotta people interested in him. He's a punch away from being ranked. But that don't mean nothing. He can't fight you with his record, he's gotta fight you with his fists. But yeah, it's on TV. Can you handle it? Just tell me if you can't and we'll back out on the deal. It's not as though we signed anything yet. We can always tuck our tails and run like scared dogs."

"I can handle it," Eugene said. "I ain't about to run nowhere."

The grim look was still on Budd's face when he said: "Kid, you'll be fine."

But Eugene was not fine, and he knew he was in over his head when he caught the first solid punch—a right-hand shot over the heart—from his opponent, Manny "Machine Gun" Mitchel, with a little over a minute gone in the first round. He had never been hit like that before, felt that kind of power. He kept his left hand working and moved as he had never moved before, backpedaling, side to side, circling left and then right. Machine Gun Mitchel, black, heavily muscled, a year older than Eugene at twenty-one, with a record of seventeen and zero, fifteen by knockout, was content to follow Eugene patiently, waiting for an opening.

On the stool after the first round, Eugene said: "Goddam, can he hit."

Budd put an ice bag on his neck and slapped him hard enough across the face to snap his head around. "Did you come here to fucking fight or talk?"

Eugene did not answer. Budd had a lot of strict rules but none more strict than that he keep his mouth shut between rounds and listen. It was that right hand over the heart that had made him forget.

". . . can't beat you because he can't catch you," Budd was saying. "You want to be somebody? You want to get out of that shipyard and live like a human being? The whole country's looking at you." He slipped the mouthpiece back in. "Don't let me down, son."

In the second round, Machine Gun Mitchel bulled Eugene in a corner and caught him with a right hook on the point of the chin, and Eugene felt every muscle in his body turn loose in instant darkness.

"Come on. Get up, goddammit." Budd's voice seemed to come from far away and Eugene opened his eyes to the bright lights of the dressing room, lying on a training table. A doctor asked him where he was and what day it was and to count backward from ten and what was fifteen take away six. Satisfied, the doctor left. Budd walked up and down beside the table spitting and hustling his balls. The back of his thick neck was very red. Eugene had not felt the punch, did not even know what he had been hit with. He tried to open his mouth, and the hinge of his jaw felt like an exposed nerve had been touched with a hot iron.

Budd turned and looked at him. He spat again as though trying to rid himself of some foul taste. "I can't fucking believe it. The first time you're tagged and you're down like a shot bird."

Eugene tried to say he was sorry, but only a strange sound came out because of the swelling in his jaw. He didn't know he was going to cry or that he was crying until he felt the tears on his face. Budd came to the table where he was still sitting with his legs dangling over the side and hugged him. Eugene felt him breathe deeply and sigh. When he spoke his voice was soft, consoling, the anger gone.

"Hey, it ain't the end of the world. So you're thirteen and

32

one instead of fourteen and 0, so what? You're the same fighter you always were."

But he was not the same fighter that he always was, and it took only the next fight back in the Beaver Street Arena in Jacksonville to prove it. An awkward, wildly swinging kid who had had only four fights caught him on the chin with a glancing blow in the third round and Eugene was out before he hit the floor. This time Budd didn't say a word in the dressing room and neither did he say anything when they got back to the apartment. A month later they went to Orlando and Eugene was knocked out again. Something had happened back in Madison Square Garden and whatever it was had taken his chin away. He knew it and he suspected that if he knew it, Budd must know it as well. He still did his road work, still ate the same good food, but Budd didn't talk to him much, and in the gym when he was working out, he couldn't help but notice Budd watching other young fighters instead of coaching him or even paying much attention. Within two weeks they were offered a fight in New Orleans.

"The purse is good," said Budd, "and this may be where you'll turn everything around."

From the way he said it, Eugene knew that Budd did not believe it. Eugene was thirteen and three now, with three straight defeats by knockout. He had been around boxing just long enough to know what he had become: the kind of fighter that other fighters could build a string of wins on. Managers anxious for their fighters to build impressive records early in their careers were willing to pay good money for a glass-jawed dog, willing even to take the short end of the purse sometimes. But Eugene was determined not to be knocked out again. Hadn't the legendary Cus D'Amato said that a fighter couldn't be knocked out unless he wanted to be knocked out? Well, he sure as hell didn't want to be knocked out again. He trained harder for this fight than he'd ever trained before, and the night before they left for New Orleans, as he sat watching

Budd at the stove making supper, he suddenly said: "I'm going to win this one, Budd. I know I'm going to win."

"Sure you'll win, kid," Budd said, but he did not even look up from the stove.

Budd didn't have much to say on the flight to New Orleans or in the dressing room or even in the corner after the fight started. Not until the end of the fourth round. It was scheduled for six and for the first four, Eugene hopelessly outpointed the boy he was going against. His jab was working and he was working the body well, and his right hand was consistently dropping over the other fighter's left.

In the corner after the fourth, Budd was beside himself. "Dammit, you've got this guy confused as a ten-dick dog! Just hold with what you got, you're doing it! You're a fighter, god-dammit! You're my fighter, I'm with you all the way. Go out and do it, Eugene. Spin him and bang him!"

Confident, full of hope, he charged off the stool for the fifth round and walked squarely into a straight right hand that landed flush on the point of his chin. He felt no pain at all, just an enormous sadness as he watched the other fighter raise his hands in the bright ring lights that were fading rapidly to darkness.

Eugene Biggs managed to get out of the ring and stagger back to the dressing room, his legs still uncertain and the world tilting everywhere about him. Budd came in behind him and slammed the door. Eugene sat on the training table, his head down. The gloves had been cut off, but his hands were still taped. He could not bear to look at Budd, whom he could hear snorting and spitting across the room.

"Listen," Eugene said, "I didn't . . ."

"I don't want to hear it! I'll tell you something else. I don't want to see you! I don't want to know you!"

"Ah, Budd, don't . . ."

"Don't tell me 'don't'! I don't wanta hear no fucking 'don't'! You the sorriest fighter ever come down the pike! Unfuckingbelievable. It's through, finished. My grandmother could knock you out."

Eugene sat numbly on the table. His whole body was cold. Even his heart felt cold, frozen in his chest. There was nothing he could say or do. He felt utterly defenseless.

"You're so fucking bad," Budd screamed, entirely out of control now, "that . . . that . . . you could knock your own self out!"

Not until he heard the door slam did Eugene look up. He was alone in the room. He sat very still for several minutes. He was not angry. He wasn't even hurt. But he was filled with self-loathing, a loathing deep and of a sort he had never before known for anybody or anything. He was also scared. Very, very scared.

He got off the table and walked to the mirror over the sink and stood watching himself. What in God's name was he to do now? He couldn't go back home, and it was not as though he were a carpenter or plumber. He had no skills at all. Nothing. He was nothing.

"You sorry sombitch," he said softly to himself in the mirror. He watched his face twist and knot as tears burst from his eyes. "That's right, go ahead and cry, you lousy bas—"

In the middle of the word, his right, taped fist shot upward and across, catching the right side of his chin. He had not known he was going to do it, and later he could not remember even thinking about such a thing. It was as if he were not involved, as if he were watching one stranger hit another. His face disappeared in the mirror, which had instantly gone black.

The ceiling swam into focus with him lying on his back. He realized what he had done. Budd had been right. He could, by God, knock himself out. He pulled himself up on the sink and looked in the mirror again. He sighed deeply and knuckled his red eyes with his taped fists. That was that. He was at the end of the road. He accepted it. He had to accept it. He was not afraid anymore. He felt burned clean, empty. His face was calm in the mirror as he thought about what had happened to him.

Budd had sold him a dream, and he had bought it, the only dream he had ever had. And now he had lost not only the dream. He felt as if he had lost his life, as well.

Chapter Three

At the curb in St. Charles Avenue Eugene looked back at the blazing lights of the mansion he had just left. The fake fight crowd was chanting something but he could not tell what it was. He bent and unlocked his helmet chained to the forks of his 1000cc BMW motorcycle. With the helmet on and buckled, he hit the start button, cracked the throttle, and popped the clutch. The front wheel leapt off the ground, and he was coming out of second gear at fifty a block away before it ever touched ground again. He laid the bike dangerously low turning south toward Magazine Street and then east toward the French Quarter. He glanced at his watch and saw that it was not quite midnight.

Pete got off from his job as projectionist at the Flesh and Flash around one-thirty, so Eugene was in no hurry. He rode past the Audubon Zoo, which was not one of his favorite places in New Orleans, but where he often went. Across Audubon Park he could see Tulane University and Loyola University where they sat cheek by jowl, darkened silhouettes against the night sky. He liked the universities even less than he liked the zoo because they reminded him of the young men and women who swarmed about them reading strange books, speaking strange languages, and studying about countries with names he couldn't even say. Since Budd had abandoned him here two years ago he had by necessity changed from a shy, quiet boy to a man who could shuck and jive the paint off the walls if he wanted to. But all the two years had really taught him was how hopelessly ignorant of real learning he was.

Charity had been responsible for that. She kept half her clothes at his place and spent half her time in it. When he passed Exposition Boulevard he saw that the lights were off in his apartment, and he knew she was not there. If she had been, she would have been waiting up for him. He didn't know whether to be disappointed or relieved.

It was a six-mile ride down to the Quarter, and on the way he stopped at an Eckerd Pharmacy and bought a bottle of Absolut vodka. That was New Orleans for you. What other city in America could you buy whiskey twenty-four hours a day, seven days a week, in a fucking drugstore, or for that matter, in a convenience store, or a grocery store? Before he got back on his bike, he opened the Absolut and took three long swallows before dropping the bottle inside his shirt. The vodka fell on his stomach unpleasantly hot, and he was reminded that he had not eaten that day. He rarely ate on a day he had to perform.

In the French Quarter he stopped at the Café du Monde across from Jackson Square. It was only three blocks to Bourbon Street where Pete worked, and he wanted to get there pretty close to the time when Pete was getting off. Otherwise, he would have to watch some dreadful thing like a girl sucking off a German shepherd or a guy being fist-fucked in the ass. Besides, if he ate something he could put a bottom in his stomach for the vodka.

He fought his way through a crowd of tourists to the only empty table he could see in the outdoor courtyard. Eugene tried to guess what sort of convention he had walked into. New Orleans was a city of conventions, and in the French Quarter he always found himself surrounded by a gang of doctors or lawyers or accountants or politicians. He was listening to two men and a woman having a drunken argument at the next table about something called postmodernism, whatever the fuck that was. He did not see George Bouchard until he appeared as if by magic at his elbow. George was wearing an apron and had a napkin on his arm.

"You check the price today?" George asked.

"We've talked about this before," said Eugene.

"I take it as my Cajun duty to try to educate ignorant crackers from Georgia. You got to get interested in kernt afeys."

"I got your current affairs swinging."

"Don't you care every time oil goes down a buck a barrel, the state of Louisiana loses fifty million dollars?"

"I try not to think about it."

"Think about this then. Till it gets back to twenty-six, I'm stuck here."

"Maybe soon," said Eugene.

"Soon sucks. If it was tomorrow, it wouldn't be soon." He paused to wipe his nose on his napkin. And then: "So how is it, Biggs?"

"I thought you'd tell me," said Eugene.

"I just did."

Eugene pointed to the tables around him. "Who?"

"Teachers' convention. The fucking Quarter's full of teachers. They don't buy nothing and they don't tip nothing. Nail down a table for two hours over one coffee and leave you with nothing but bullshit hanging in the air." George looked off into the middle distance. "Did you hear the Japs wanted to buy Louisiana but the Arabs wouldn't sell?"

Eugene said: "That was just a little funny the first ten times I heard it."

"Yeah, well," said George. "The goddam Arabs took this fucking state and never fired a shot."

George had been a welder on an offshore rig until the Arabs had driven down the price of oil from thirty-six dollars a barrel to five dollars a barrel. The whole state, which ran on oil money, had gone belly-up, and George had to give up his torch for tips. It had left him in a state of perpetual outrage.

"Jesus, can you imagine a Cajun wearing a fucking apron and bringing coffee to a teacher?"

"Luck of the draw," said Eugene. "Besides, I ain't a teacher. Bring me beignets and a big milk."

George was gone for what seemed a long time but finally came ambling back with the order.

"Damn, you slow," said Eugene.

Over his shoulder as he walked away, George said: "I don't make enough to be fast."

Eugene ate the beignets slowly, chewing a long time, and sipped the milk. Maybe he'd get a steak with Pete later, or maybe he would wait until he got home and make himself something. For now the cold, sweet milk soothed his outraged stomach, and the beignets—tiny and sweet and so freshly made they were still warm—always made him feel good because they reminded him of his mother's kitchen on Sunday. She would bake something—a pound cake or apple turnovers or a blackberry cobbler—before they left for church in the pickup, and the house would be full of the smell of it when they sat down to eat in the early afternoon.

He swallowed the last bite of the last beignet, drained his milk, and walked toward the heart of the Quarter, Bourbon Street. In one of the first letters he had written from New

Orleans, he had told his daddy about eating a beignet and told him you had to say it "ben-yea," a donut that was square instead of round and that had no hole in it. His daddy had written back: "How can it be a donut if it ain't a donut. I believe I don't understand and I believe them ain't our kind of people." Eugene smiled. There was not much about New Orleans his daddy would understand. He wouldn't understand people staggering about the sidewalks with go-cups of whiskey in their hands or paying a hundred and fifty dollars for a meal or watching two naked women lying on the floor, one on top of the other, humping away while the audience howled.

This part of the Quarter was closed to automobile traffic, and Bourbon Street was jammed from curb to curb with sweating tourists, some shouting or singing, some dancing to music only they could hear. You'd never know, he thought, that New Orleans was in the midst of a depression, swimming in a sea of red ink, from all these people celebrating God knows what here in the middle of the night. But he had long known that New Orleans was a circus, a circus that didn't travel, but a circus nonetheless. And people always went to the circus, in good times and bad. He passed three young black men taking turns break-dancing on what appeared to be a quilt. Tourists threw money at them, and the young black men smiled for all they were worth.

Eugene turned off Bourbon into an alley that dead-ended at the Flesh and Flash theater. The men and women trying to buy tickets were too drunk to line up, but simply milled about, always pressing forward, their money in their hands. They wore straw hats with red labels on them that said: TEACHERS ARE PRECIOUS.

He went straight to the door where Billy Powell took tickets and yawned behind his hand.

"You any good tonight?" said Eugene.

"I'm good every night," Billy said without enthusiasm.

"How long till the show's over?"

41

"How should I know? I'm fucking dying here."

Eugene took the stairs two at a time up to the little room that was the projection booth and let himself in. Pete sat in a folding chair in front of an open window that looked out over the audience. A babble of Spanish filled the theater. Porn with subtitles. Jesus. There was no air-conditioning and the room was heavy with the sour smell of sweat and alcohol and urine floating up out of the theater. Pete looked over at Eugene and his teeth flashed in his black face.

"I'd kill for a drink," said Pete.

Eugene nodded toward the window and the people below. "It's enough alcohol down there to keep you drunk for the rest of your life."

"It's down there and I'm up here."

An agonized groan came out of the audience.

"I thought you might need a taste," said Eugene, pulling the bottle out of his shirt.

"Thank God," Pete said, taking the cap off. His hands trembled slightly.

"This'll calm me down." He turned the bottle up, took off half of the vodka, and wiped his mouth. "I sent Billy for a fifth but I finished that before the shift was half over."

"You mean you already . . ."

But gasps and cries of real pain from the audience stopped him and he glanced toward the screen. Two Mexican bandits, swarthy, with greasy black hair, knives sheathed on their belts, stood on either side of a naked young man tied to a chair. The young man had safety pins through his lower lip and the bandits were cutting his ears off. Slowly. The young man and the chair were covered in blood. There was not an actor in the world who was good enough to do what the young man in the chair was doing, to make the sounds he was making.

"Sweet fucking baby Jesus," said Eugene, "you got another one."

"Yeah. It's been packing'm in all day. Snuff films always a

great draw. You wonder why I got through a fifth in less than four hours?"

"Not anymore I don't."

"I watched that poor fucker butchered alive six times today."

"Then don't watch."

"It's hard not to, man."

"I know," said Eugene, looking away as one of the Mexican bandits was going into the young man's mouth with a pair of pliers. He went over to the far wall where he could not see the film even if he wanted to. Pete turned in his chair, his back to the screen. The audience had stopped groaning and gasping and was now cheering as if for a racehorse or a football team.

Pete took another pull at the vodka bottle. "You know Tulip?"

"The new girl at Maxie's?" said Eugene. "The one that does the thing with the teddy bear?"

"She went home with me last night," Pete said. "It was like throwing a starving man a turkey."

"What'd it cost you?"

"Cost? Man, you know I don't pay."

"How much?"

"She did it for love," Pete said. The wild cheering down below swelled and rolled. Pete glanced back toward the screen. "You ought to watch this. They're about to kill the guy. You really ought to see it."

"I don't think so," said Eugene.

"B and B was asking for you," said Pete, his back still to Eugene, his voice soft, abstracted by whatever was happening down on the screen.

B and B stood for Bad and Beautiful. Bobby "Bad and Beautiful" Barfield was a ranked heavyweight from New Orleans but trained and fought out of Baton Rouge.

"How's his hands?" said Eugene.

"He ain't got no more trouble with that."

"He says."

"Yeah," said Pete, "that's what he says."

Eugene saw Pete's back start to quiver and his fists clench. The audience screamed as a single voice and then went silent. The door to the projection booth opened and an old man came in wearing Bermuda shorts and a red T-shirt with an aluminum horse embossed on the front of it. His name was Rankin Shortly and he was the projectionist on the next shift. Pete spun from the open window looking down on the screen. A forked vein stood in his forehead.

"It's about time, Rank, you brainless old motherfucker."

"What the hell's chewing on you?" Rankin looked at his watch. "I'm right on time."

"Want me cheerful? Try being early once in a while."

Eugene followed Pete down to the street. They walked toward the Café du Monde, where Eugene had parked his bike.

"You go over to Baton Rouge to see B and B?"

"Naw. He was over here. Just goofing and come by to see me. You know what he wanted? Wanted to know if I'd work as a sparring partner. He's got that big fight coming up with Cobra Carnes in three weeks, thought I could help him, give him some speed work. He knows how I fight."

"Used to fight. You don't fight no more, remember. I fucking hope you gave him the right answer. All you need is for the Bad Man to pound you five or six days in a row."

"He offered me fifteen bucks a round, man. Fifteen bucks ain't bad money for three minutes' work."

Eugene watched a man in a white linen suit playing a trumpet in the middle of the street. The tourists pitched pennies and nickels, dimes and quarters at the trumpet player's upturned Panama hat lying in front of him. They seemed more interested in hitting the hat than they did in the music.

"I hope you gave him the right answer."

Pete caught Eugene's arm. "Hey, you take that gig tonight?"

"Yeah," said Eugene, jerking his arm free, "I took it."

"And you played it?"

"I always play it if I take it. I don't fade on nobody."

"Man, fuck! What is that? Seventy-some-odd times?"

"Seventy-three," said Eugene.

"And you telling me about right answers?"

Eugene shrugged: "It's a lid for every pot."

"You got the wrong lid for yours, my man. Believe it."

They were at the motorcycle, and Eugene unlocked the spare helmet from the seat bar and handed it to Pete. "Where you want to eat? I'm hungry."

"How bout Thibodaux's?"

"Good enough."

Eugene swung his leg over the bike and hit the start button. Pete put his hand on Eugene's shoulder. "Sorry for the heat about the gig. It ain't nothing," said Pete.

"Right," said Eugene, roaring out into traffic. "It ain't nothing."

Eugene's bond to Pete was boxing, and he felt closer to him than he had ever felt to his own brother. Pete had been the fighter who had knocked him out that last time, the knockout that had caused Budd to abandon him right here in New Orleans. Fourteen months later, Pete had been ranked seventh in the world, a spectacular rise, but not uncommon in boxing. And six months after being ranked seventh, he was the projectionist at the Flesh and Flash, a spectacular fall, but not uncommon in boxing.

At Thibodaux's, a wooden structure on the Mississippi River that resembled a warehouse more than it did a restaurant, both of them ordered crab cakes, red beans and rice, and shrimp etouffée.

"And I'll have a double vodka straight up, with a half a lemon on the side," Pete said.

"Working pretty strong to the bottle, ain't you," Eugene said, ordering iced tea.

"I'm heavy with thoughts, man. Besides, I ain't driving."

Pete ordered two more drinks before the food came and when the food did get to their table he never touched it, but kept drinking while Eugene ate. Pete sipped steadily at the vodka and watched through the window as huge freighters steamed up the Mississippi.

"Wanta know the truth," said Pete, "she woulda never sat down at the table if B and B hadn't been with me."

"Who you talking about?" Eugene said, but he already knew. And he also knew that Pete was as drunk as he had ever seen him.

"Tulip," Pete said.

"Bobby Barfield is bad and he is beautiful," Eugene said. "He's also a world-ranked fighter with money in his pocket. What do you care why she sat down at the table as long as she ended up back in your room where you could strap her on for a quick thrash."

"It wasn't no quick thrash." Pete raised his hand and the waitress brought him another drink. "She was with me six hours." He held up his hand, fingers spread. "Count'm: one, two, three, four, five, six."

"You better lighten up on that booze. You go down, I'm leaving you where you fall."

Pete said: "No, you wouldn't."

Eugene knew he wouldn't, but he said: "Try me."

"I don't want any more anyway. I got things to do tonight."

"Good." Then: "You want that food?"

"I thought I did but I don't."

"Good again." Eugene shoved his own plate away and took Pete's.

Pete looked out at a passing freighter and said: "You ever wish you could git on one of them things and just float away from everything?"

Eugene kept forking the etouffée into his mouth and watched the freighter, too. He swallowed and sucked his teeth.

"You and me ain't sailing nowhere, Buckshot. We got both feet nailed to the floor and a anchor in our ass."

"You talking about yourself. You ain't talking about me."

"I'm talking about both of us."

"You got Charity. I ain't never had myself nobody like that. I got to get me a squeeze like that. I had a squeeze like Charity, I'd have a anchor in my ass too. I just wish I had a woman like you do."

"No you don't."

"I got eyes, man. I seen Miss Charity."

"Yeah, you got eyes."

"You strange, Eugene. You don't . . . you don't signify, man."

"Don't worry about it. Watching that sucker get killed all night's done ruint you mind. You'll feel better tomorrow when you head is reminding you of all the vodka you drunk up tonight."

Somebody put money in the jukebox and Doug Kershaw's Cajun fiddle sawed through the restaurant. Pete chewed at the lemon half and when the music died he said: "You was right. I paid her."

Eugene was finishing the last of his beans and rice. He was not interested in talking about Tulip.

"She wanted a hundred bucks," Pete said, "and I give it to her."

Eugene put his fork down. "You give her a hundred bucks." It was a statement, not a question. "For Christ's sake, Pete, what did you want her to do?"

"Fall in love with me," said Pete.

Eugene smiled. "So you and her gonna get married or what?"

But Pete was lost somewhere else or maybe he was so drunk he did not hear, and Eugene was glad for that because he was sorry about what he'd said.

"It was something else, man, unbelievable," Pete said. "I

told her I wanted her to be in love with me for the rest of the time she was there."

Eugene did not want to hear this. But he knew Pete wanted to tell it. He dipped a lemon peel out of his tea and chewed on it while he watched another freighter churn past. Pete drained the last of the vodka and set the glass back on the table carefully.

"She said for a hundred dollars she could fall in love with me till nine o'clock."

"Jesus," said Eugene.

"So I put the hundred down, and she set the alarm clock for nine."

"Jesus," said Eugene.

"And she was in love with me. Really in love with me. I never had . . . I never . . . it was never like that in my life."

"Jesus," said Eugene.

"And then the alarm went off at nine o'clock and it was gone. She was gone."

Eugene thought that he had to get different friends or a different life or else get meaner. He wasn't tough enough for shit like this. "I need a drink before we go," he said.

Pete looked at his watch. "We got time. I might as well have one with you. My brain is wet, man. One more won't hurt."

The waitress brought the drinks and Eugene asked for the bill. "On me tonight," he said. He brought the folded bills out of his pocket.

"Eugene, you shouldn't spend that money on me," Pete said. "I don't feel right eating and drinking on that money."

Eugene felt a hot rush of anger but it was not directed toward Pete. He didn't know what or who the anger was for, but it was anger just the same. "What is it with this money bullshit? What kinda money we talking about?"

Pete put his drink on the table and leaned forward. "It don't matter who knocks you out, Eugene. A knockout is a knockout.

It ain't two cents' worth of difference if somebody else does it or if you do it yourself."

"You getting to be a huge pain in the ass," said Eugene. "You know that, Pete?"

"I know what I know."

"You don't know nothing about my life, you just think you do." He knew he ought to ignore Pete. He was drunk and you might as well talk to a fence post as talk to a drunk.

"You sell me short, boss man. Nobody knows nothing about nobody. Can I tell you something? Thing I heard on the radio. Lemme tell you."

"You're just boring the shit out of me, Pete."

"This ain't gone bore you. This is . . . well." He finished his drink. "It's a toymaker putting out this new kind of toys. Heard it on the radio. It's supposed to make kids that been hit with bad shit feel good about the bad shit. I don't know myself, but that's what the man on the radio say."

"Pete . . ."

"Dolls. Fucked-up dolls. They got a little blind doll. She come with a walking stick painted red on the bottom and a Seeing Eye puppy. Dig? Got a little boy doll tricked out in running shorts and running shoes, but he got braces on his legs. No, wait. Lemme finish. It's more dolls. But you got the picture. Now, they got one doll you can buy that ain't got nothing wrong with him, but you can buy this shit on the side to make him fucked up any way you want. You can buy a blind walking stick and put it in his hand or you can buy a little wheelchair and put him in it or braces for his legs or hearing aids. All kinda shit to fuck him up like you want him. Now, dig this. You oughta buy that little doll that ain't got nothing wrong with him. The doll that ain't got nothing wrong with him, see, that's you. Then you can fix him up like you gone be if you keep on knocking yourself out. Blind? Wheelchair? Mouth won't talk,

49

legs won't walk. Truly, one fucked-up boy. What do you think, man?"

"I think if you was sober, I'd slap the shit out of you."

Pete smiled and shook his head. "No you wouldn't. I may be half in the bag and such a weak stick that I couldn't help watching a guy get killed six times today, but you ain't gone slap me, and you ain't gone leave me where I fall either, because I love you, man."

"Let's get out of here," Eugene said.

Pete glanced at his watch. "You right about that. I got to get back to the Quarter."

"The Quarter? You not going home?"

"After I go to the Quarter. Tulip gets off at three."

"They got a doll without any balls?"

"Don't."

"Because Tulip's got yours in her purse."

"I wish you wouldn't talk like that about Tulip. Give me a shoulder. I've come down with a case of the dark twirlies." Pete threw his arm across Eugene's shoulders.

"You're too drunk to fuck," Eugene said.

"But I ain't too drunk to be in love."

"I guess not."

When he got to the Quarter, he ignored the barriers that had been set up to prohibit motor traffic and roared right onto Bourbon Street, scattering drunken tourists like a covey of quail, and stopped in front of Maxie's. Pete got off, unbuckled the helmet, and watched Eugene lock it on to the bike.

Eugene laid his palm against Pete's cheek: "Slide with it, champ, just slide with it."

"Nother time, nother place, Knockout."

"Don't call me that."

Pete pointed his forefinger and cocked his thumb. "The day you quit, Knockout, the day you quit."

Eugene jerked his bike onto the back wheel and blasted down Bourbon Street, narrowly missing several graying women in a conga line. The white T-shirts they were wearing had red legends, front and back, that read: TEACHERS ARE PRECIOUS.

Eugene did not realize how exhausted he was until he was pushing the bike into the garage behind his apartment. He had slept late that morning and had just eaten a good meal. There was no reason for the bone-tiredness he felt. Maybe it was the alcohol. He drank only on the days when he performed, and then only two or three ounces to take the edge off his nerves. The apartment was still dark so he did not have to check the street for Charity's yellow Cadillac, a convertible, but he did anyway and was relieved when he did not see it. He opened the back door and walked into the kitchen, so chilled from the central air-conditioning that it was like diving into cold water. It was a shotgun apartment, with a small dining room directly in front of the kitchen and then the bedroom followed by a short hallway, off which was a bath. The hallway led into a combination den and study and beyond that was the living room. The front windows of the living room looked directly onto Audubon Park. A park filled with ponds and shrubs and trees and gazebos and a golf course and jogging trails and biking paths. The apartment was painted entirely in off-white with gold curtains at the windows. The hardwood floors were bare and gleaming except for occasional Oriental rugs which Charity had thrown about with careful abandon. Eugene slipped out of his clothes and into a robe. He pulled the covers back on the four-poster double bed and stood regarding it a moment before walking into the combination den and study where several hundred books sat in shelves surrounding an area that Charity called the pit, which was a horseshoe-shaped couch facing a huge television set. Cater-cornered from the couch was a word processor on an oaken table. Beside the table on the floor sat a two-drawer steel cabinet with a heavy padlock on it. It held the notes for the book Charity was writing. It was fireproof, she said, and it looked like it would take dynamite to get through the lock.

"A burglar is not gone steal your notes, Charity. You caint hock notes. Nobody cares about them."

"Maybe not now," she said, "but someday everybody will care about them."

He let it alone. She was a passing strange girl, and he had other things to occupy his mind.

He sat down at the word processor and put his fingers on the dead keys. In front of him, ranged on the table behind the word processor, were books, thick books with raised titles.

He read them aloud, his voice barely above a whisper: *Interpretation of Dreams, Abnormal Psychology, Aberrance, and Beyond, Civilization and Its Discontents.*

Eugene had never read any of the books. They did not belong to him. Neither did the word processor. He couldn't even type. The books and everything on the desk, everything in the apartment except his clothes, belonged to Charity. But he liked to sit at the desk and pretend that all of it was his, pretend that the apartment—by far the grandest place he had ever lived in—was his. But the apartment was Charity's, too, leased in her name, although she did not stay there. She called the apartment her living laboratory.

"What the fuck does that mean?" he had asked.

"It means you and your friends are a walking gold mine," she said.

"What does that mean?"

"It means I love you." And she would put her great wet mouth over his and then gradually move up his face, kissing slowly, and end by licking the scar tissue in his eyebrows.

"Don't do that," he would say.

"But I love to do that."

"I worry sometimes that you might be just a shade crazy."

"Crazy? A meaningless term. I don't know what crazy is. Do you know what crazy is?"

"You damn right I do."

Eugene took his fingers from the keys and thought of the hundred-dollar bills folded in his pants pocket. He ought to write his daddy a letter but tomorrow would be soon enough. He couldn't mail the money home until the morning anyway.

Chapter Four

Eugene sat on a bench in Audubon Park and read his daddy's last letter. It was written in pencil on ruled tablet paper.

Deer Eugene
Your brother broke his arm an is off work til it is heeled up if it ever does and its been no rain here for so long I bout forgot what it looks like. the crops is burning up on the feeled an you ma has been sick in bed for near bout 2 weeks with the female trouble. 4 of the calfs got the scours an two of them have shit there selfs to death. eugene we are all of us fine an hope you are the same. son i know your uncle carter and me was purty hard on you when you tuck up the prize fighting but i guess you showed us and the whole fambly is so proud and glad for the money you have sent to hep us out. I wish more than anything i could see you in one of them boxing fights but i prechate your feelings not to want me or your ma or any

blood kin to see you in one of them fights. i believe myself if it was
me i would not want blood kin to see me do it if i was doin it as
you are doin it. i never seen a boxing fight but i seen plenty of fights
when i was young mebbe that is where you git your fighting from
me. mebbe it will rain soon an you brothers arm will heel up if it
ever does and your ma will git over the female trouble but i guess
them two calfs that shit theirselfs to death is gone forever.

love daddy.

Eugene turned the letter in his hands and watched the men
and women passing twenty yards in front of him on the jogging
trail. It had rained earlier in the morning and it was probably
only about 95 degrees now. There was not a cloud in the sky
and not the slightest breath of wind.

A herd of mongoloid children, flanked on either side by their
attendants, galloped past on the jogging trail. He watched the
children until they disappeared into the trees and then he
picked up the writing tablet he had brought from the apartment and propped it against his knee.

Dear mama and daddy and Edsel
I have been working with afflicted children hear in Audubon park
which is a park hear in New Orleans. Being a athalete you get
asked to show up at functions and things like working with afflicted
children. I wish . . .

What he wished was that he could write better, that he could
get what he wanted to say down on paper in a way that was
not so crippled. And more than that, he wished he did not lie
to his daddy and his mama and his brother, Edsel. He did not
even have to lie. He knew that. But there was something that
made him want to lie, or at least made lying necessary. It was
all very confusing. He knew it wasn't really necessary to send
money home. He could just stop. Except he felt he could never
stop. His daddy and mama and Edsel needed the money badly
and he had got used to being able to help them out. Writing

the letters and sending the money made it easier to do what he was doing even if he hated what he was doing. But he had to keep doing what he hated so he could write the letters and send the money. He shook his head violently, as if to clear it after taking a punch. These were strange times. Maybe Pete was right, maybe he had knocked himself out too many times. He looked back at the last two words he had written and could not remember what he wished.

"I thought I might find you out here."

Charity came around the bench to stand in front of him. She was wearing a bright yellow jumpsuit that didn't have a wrinkle in it. Her clothes never had wrinkles and her breath never smelled bad and her hair was always permanently in place.

"I'm writing a letter," he said.

She sat beside him and placed one of her wonderfully manicured, immaculate hands on his knee. "I'm glad," she said, showing teeth so perfect they should have been false.

"Don't be glad," he said.

"Then you worked last night?"

"Yes."

"How did it go?"

"Charity, I'm trying to write a letter."

"We've all got to work, darling. Don't be so touchy about it. There is work and there is work." She pointed her feet and examined her carefully painted toenails where they showed square-cut and lavender at the end of her sandals. "You'll be happy to know my paper was well received."

"Your paper?"

"The one I gave." The glistening point of her tongue darted over her lips. "You sweetheart, you forgot. There's a teachers' convention in town. They're staying at The World Under Hotel in the Quarter. That's where I gave the paper." Her tongue made a little noise against the roof of her mouth. "And you forgot!"

"I did, I forgot," he said.

"I'm disappointed," she said. But she didn't look disappointed. "I wish my work meant as much to you as your work means to me."

"I don't know anything about your work," he said. "I'll never know anything about it."

"Of course you will," she said. "You have your whole life in front of you. We both have our whole lives in front of us."

He looked at her. "To have an education, you sure do say dumb things sometimes."

"I don't see what's dumb about having your whole life in front of you."

"Did you ever think that the guy that dropped dead five minutes ago had his whole life ahead of him? Jesus! My life, your life, may be one more day or one more hour or one more case of cancer or one more . . ."

Her hand dipped toward the pocket of her jumpsuit and came out with a little pad. A tiny gold pen was clipped to the front of it. She wrote something quickly and then said: "Or one more what, Knocker?"

"Eugene, goddammit. Eugene," he cried, loud enough to turn the head of a passing jogger.

"You haven't been a Eugene in a year and a half. You're a Knocker now. It's what makes you special. I don't see why you're offended."

He watched the herd of defective children galloping back along the path. "I may be a Knocker, but my name is Eugene."

"You want to tell me about your night?"

"Not now."

"Well, I had a good night." She crossed her long, slender legs. "Teachers *are* precious." The thinnest of gold chains made of perfect tiny links moved on her wrist as she absently stroked his leg. "That's the theme of the convention in the Quarter. It was a big honor, a really big honor, giving the paper, I mean. It's rare for an ABD to be asked—or allowed, for that matter—to give a paper at a convention like this one."

ABD. Eugene had heard those letters from the beginning. It was a long time before he knew that they stood for All But Dissertation. And longer still before he came to realize what a dissertation was. Only recently did he find out that he, Eugene Biggs, was the subject of her dissertation.

you and mama and edsel should see all the people I get to meet being a boxer. Even teachers. It is one teacher who . . .

He sighed, ripped the paper out of his tablet, and crumpled it. He would write the letter later. There were limits to what even his daddy would believe.

Charity looked at the ball of paper he had in his hand. "Now why did you do that? Tell Charity what's the matter. Something happened last night, didn't it?" Her fingers moved nervously on the pad.

"Nothing happened last night that ain't happened before or won't happen again."

"You must not think like that," she said. "Nothing has ever happened before and anything that has happened will never happen again. The discrete experience is nothing; the theory that holds it together is everything."

"You make my head hurt when you talk like that."

"So you've told me," she said. She placed her smooth palm on the back of his neck. Her long cool fingers gripped and probed with surprising strength. "Look at me, Eugene." He turned his face to her but kept his gaze averted. "Please don't," she said. "Don't treat Charity like that. Look into her eyes." He did and her eyes were deep green with gold specks in them. The cool hand on the back of his neck suddenly went hot. He felt his whole body flush. "Don't you feel good when you're with me?"

"Yes."

"But you sometimes feel bad too," she said, "very bad."

"Yes."

"You monosyllabic sweetheart, I love you."

He did not know what the word *monosyllabic* meant, but it didn't matter. He could feel the heat coming off her incredibly long, incredibly smooth body.

"You want to tell me many, many things, don't you, Eugene?"

"Yes."

"But the things you most want to say, you have never been able to say, have you?" When he didn't answer, she raised her hand and traced the scar tissue in his eyebrows and drew her fingers over his nose and mouth, stopping at the point of his chin. "I know," she said. "But you will, I promise you, you will be able to say all of it, everything, to me someday."

"I don't want to tell you everything."

"You will. Trust me, you will." She got up and walked a little way from the bench and stood watching him. A tiny pulse beat in her neck. Her hair, neither blond nor red, but something in between, glowed like a light. "What time did you get home?"

"I don't know. Late."

"Did you go to the Flesh and Flash?"

"Yes, and before you ask, the answer is no."

"No, what? You didn't even ask him."

"I don't think Pete likes you."

She smiled and shook her head. Her shining hair flew about her face and then fell, every strand of it, back into place. "Everybody likes me."

"Maybe Pete doesn't."

"Maybe you didn't talk to him. All I'm doing is extending a dinner invitation. He could bring a girl. We'd have a good time. That's all I want."

"I know what you want," he said.

He was sorry he had ever told her about Pete and sorrier still that he had ever introduced her to him.

"I find him fascinating. I never knew a ranked fighter before and he was your last . . . connection, so to speak . . . with the

59

world of boxing." She came back to the bench and stood in front of him and put her hands on his shoulders. Her taut stomach pressed against his face and the deep rich smell of her came into his nostrils. He watched his hands rise and take the inward curve of her hips, felt the pulse in his thumbs beat against bone. It was not what he meant to do or even wanted to do, but wanting had nothing to do with it. "Now," she said, her voice almost a whisper, softly exhaling the word, "I'm going back into the apartment. I'll wait for you. You will not keep me waiting long. You will not." Her hands moved to his face and turned it up toward her. She bent to him—he lost in her green, smiling eyes—and her lips just brushed his before she was gone.

He turned on the bench and watched her move over the grass toward the dappled shade under the arching limbs of the enormous oak trees on the edge of Audubon Park. She walked the way she did everything else, with great strength and precision, as though her movements were calibrated to certain very fine tolerances. It was the first thing he had noticed about her. . . .

When he came through the swinging kitchen doors of Jacques' on Royal Street, balancing two platters of steaming oysters garnished in a butter sauce, Eugene saw from all the way across the crowded restaurant precisely the measured way her upper lip moved to meet the lower one, and further he saw her gold-flecked eyes measure him as he hung suspended in the instant. And then the instant was gone. He hustled the oysters to a table against the wall farthest from where she sat, his lean boxer's hips moving among the chairs where young men and women laughed and drank and sucked their buttery fingers.

And he continued to see her all night—the way her gestures moved to some carefully calibrated secret of their own—stealing little glances at her, and seeing at the same time, or so he thought, that she was holding him with her own eyes, little delicate calipers, that never quite let him escape. But he never expected to see her again after that night. Why should he? Wasn't the French Quarter always filled with young women trailing an effluvium of musk on the air, strutting their perfect bodies which were the result of having been raised on the simple best? How might one of these young women seek him out from all the rest and bestow upon him her strange, mysterious favors? It was the stuff of every horny boy's secret dreams.

But it happened nonetheless, and happened in about as straightforward a manner as could be imagined. She came into Jacques' the next night and walked straight across the restaurant to him, moving with that glass-smooth stride of hers, the heft of her breasts perfectly divided and ticking counterpoint to each step she took.

"Which is your table?" she asked.

"What?" Her breasts were still ticking in his eyes and he couldn't think.

"I want to eat at your table. I want you to bring me my food."

There was something deliciously obscene not so much in what she said as the way she said it, the way it came out of her mouth, that and the way she looked out of her eyes.

"My station's those four tables." He pointed. "There, against the far wall."

She went directly to the table nearest the window and sat down. Her manner, he would soon learn, was always to be direct. It seemed a necessary extension of the unnerving exactitude of the way she moved.

When he came to her table to take her order, she said: "I'll have blackened redfish, and this nice little bottle of white wine." Her long, finely shaped finger underlined the bottle on the wine list. "And followed, I think, by black coffee and a cointreau." And then in the same breath: "What time are you finished here?"

"You mean get off?" He was startled by the question.

She smiled, but when she spoke, her words were clipped, precisely spoken, direct. "I mean, what time are you finished here?"

"Two-thirty."

"Good. I thought I'd like to know you. I thought a drive out Elysian Fields to Gentilly might be nice. My car's in the parking lot. I might leave, but if I do, I shall be back at precisely," and the syllables of the word ticked off her tongue like pebbles dropping from her mouth, "precisely at two-thirty. Will that be sufficient?"

He could only nod, which infuriated him, but her manner had taken his voice.

And she did reappear precisely at two-thirty, but they did not drive out Elysian Fields to Gentilly. They drove instead to Napoleon between St. Charles Avenue and Magazine Street. When she stopped in front of a house that was a huge pile of marble, she simply said, "I live here." And got out.

When they got inside she splashed something into a glass for herself and then asked if he would like a drink.

"I don't drink," he said, and he did not in those days, had never tasted anything alcoholic.

"You don't drink? Everybody drinks."

"Not fighters."

"Fighters?"

He enjoyed that. He had broken through to something human, something he could understand.

"I'm a boxer. Well, I was a boxer," he said.

"Tell me about it. Really, I'm genuinely interested."

And he did. He told her about meeting Budd, about his incredible winning streak, and finally about the disastrous fight in Madison Square Garden. Nothing had ever worked right after that, he told her. But he couldn't bring himself to tell her about Budd abandoning him here in New Orleans.

That did not seem to matter to her, though. She put her

drink down and came to him. "Oh, you poor baby. What an absolutely fascinating, wonderful"—she kissed him—"case. This must be my lucky night. I want to know everything about what makes you . . . tick."

Eugene Biggs did not understand anything about the way she was talking. Then things went from confusing to terrifying. Because she started shedding her clothes. She came out of them with a certain mind-numbing precision. And as each piece of clothing fell away, she was revealed to him more and more beautiful: taut, symmetrical, muscled long and lean and flat like a gymnast or a sprinter. Until she was finally down to a pair of panties so brief they didn't have half enough cloth in them to make a glove.

Eugene could only think of the thirteen whores Budd had bought him—had insisted on buying him—and that those cheap thirteen, taken briefly in the dark, were all he knew of women. And certainly none of them had looked like this woman—this Charity, as he had learned she was called.

"Wait," he said. "Don't."

"Don't what?" she said.

"Just wait," said Eugene. "I want to show you something."

She smiled. "But that's what I'm trying to do for you, you sweetheart, show you something."

"I know," he said. "But I can do something you've never seen before, never even heard of. I can do a . . . trick."

She dropped her head to one side and regarded him for a long moment. Then in a strange, quiet voice, she said: "You wondrously peculiar boy, what have I found in you?" Then after a moment she said: "All right, go ahead. Do your trick for me."

She was standing naked except for the flesh-colored panties. Her fists were on her hips and she stood with her legs apart, her chin thrust forward. She was obviously waiting for him to prove himself, and obviously expecting that he would make a fool of himself. Well, he'd show her.

He shook his shoulders to loosen them, set his feet, and let his jaw go slack, unhinged. He snorted once and caught himself on the point of the chin with a vicious right cross. He heard her gasp and say what he thought was "Oh, God!" before the lights went out.

Her inconstant voice came swimming to him and he realized he was lying on the bed with her face above him. She had a penlight in his right eye.

"Subject definitely comatose. Subject rendered senseless by blow to chin with own right fist. Eyes dilated. Breathing shallow."

Her face focused above him and he watched her reach to hit the button of the Dictaphone she had been speaking into.

"That was only sensational," she said.

He shook his head. The grogginess was going. "I told you I could do a trick."

"That was no trick. That was truly an incredible feat. You don't realize what you've got there. It could make you famous." She looked off toward the wall and her eyes took on a thousand-yard stare. "But just as importantly, it might make me famous as well."

"I don't understand."

She reached down and touched him and he realized his clothes were gone. She had undressed him. She brushed his cheek with her lips.

"We'll put off the fun and games for a moment," she said, "and I'll explain a few things to you. I'm working on a PhD here in New Orleans and I'm ABD." She saw he did not understand. "I'm working on a doctorate in psychology and I've done it all—or I will have done it all in a couple of months—except write the dissertation. Don't look like that. It just means a book. I have to write one and I've been searching for something. I have certain . . . interests. And I think Providence must have sent you to me. Understand, you are a metaphor for most of what I believe about the world. You're looking at me like

64

that again. All I'm saying is most people are standing around waiting to see somebody knock himself out—in a manner of speaking. It's human nature. Not pretty, but true. Listen, did you ever hear or read about somebody standing on an eighth-floor ledge of a building in New York City? And who's down in the street? A huge crowd of people. And what are they shouting, darling mine? They're shouting *Jump, jump, jump.* They want to see the poor bastard do the ultimate knockout on himself. Do you understand?"

"Sort of."

"I've probably told you more than you really need or want to know. But there is one more thing. It's about the theory."

"The theory?" he said.

"The theory I want to make," she said. "Discrete"—and she saw his look—"individual, separate facts are meaningless. But if you could make a theory that accounted for all facts, that is, make every fact related to every other fact in a necessary way, it would make your name forever." She sighed. "All right, let me put it another way. I'm going to construct a theory that will relate in a necessary and inevitable way why you pick your nose to why you scratch your ass."

"I don't pick my nose and I don't scratch my ass," he said.

"God bless you," she said. "You're blushing. Well, never mind, you'll understand by and by . . . more or less." She rubbed his stomach. "You've got a beautiful body."

"I'm a fighter," he said. "I'm supposed to have a beautiful body. A fighter, dammit."

"Sure you are," she said. "Sure you are."

And then her panties were gone and she came to him so quietly, so tenderly that when it was over and he lay beside her sweating as though he had just come from a fight in the ring he thought that this must be what love was. He had long wondered about love. He knew he loved his mother and father and brother and that they loved him. He understood that. But love between a man and a woman was a mystery, had always

65

been a mystery. He had never seen his father so much as hug his mother. But obviously somewhere, somehow, they had been close enough or else he and Edsel wouldn't be in the world.

Now, lying beside Charity, he felt he had made a discovery that would change his life, make him see the world in a different way.

"Where do you live?" she asked, tracing a little design on his chest with her slender fingers.

"I got a room in the Hotel Rue."

"That's a rat's nest."

"It's all right," he said.

"I don't want you there."

"It ain't a choice. Never was."

"There is now," she said. "I know a place on Exposition Boulevard. Right on Audubon Park. You'll love it. I'll get it for you."

He raised up on one elbow, astonished. "Why would you do that?"

Her voice was calm as she ticked off the reasons. "One, I can afford it. Two, I want to see you again, because, three, I find you interesting and potentially valuable to my work. And four, I am not about to set foot in the Hotel Rue. It's bad enough to drive by it."

"Listen, I don't know what you think I am, but I ain't a fool. What do I have to do for this place, rob a bank or some damn thing?"

"Be my friend . . . and my lover."

"Lover? You mean fuck you?"

"Fuck me, yes, if you prefer to say it that way."

"Where I come from we don't know nothing about being no lover. We got fucking and we got marriage."

"Marriage is outdated."

"Outdated. What outdated?"

"Over. Finished. Done with. Kaput."

"Don't tell my mama and daddy that."

"I won't, don't worry. Where do you come from, anyway?"

"Bacon County, Georgia."

"My God," she said. "This gets better and better."

"No shit!" He was angry, really hot. "Where you from?"

"Dallas, Texas."

"A fucking foreign country," he said.

"Don't knock it, son. This is oil money we lying up on here."

"I thought the shit had been knocked out of oil and everybody in it."

"Men like my father never play with their own money. He's got enough stashed to keep me in Cadillacs forever. He's happy to sponsor my research, and before you ask, that just means what I'm examining, looking into."

"And what are you looking into?"

"You, sweet man-child, you." And then she turned to him again and her tenderness and the soft crooning of her voice smothered all his anger.

Eugene had his doubts about staying in an apartment Charity was paying for. None of it linked up very well with what had happened to him since he left home. Your best friend in the world, your fight manager who was like a daddy to you, abandons you in a strange city where you are reduced to carrying food to strangers, bowing and scraping for tips, and then a beautiful, rich girl comes along, fucks you, treats you like kin, and on top of it offers to rent an apartment for you for no other reason than you interest her, wants to be your friend and offers to let you keep on fucking her to boot. It did not make sense.

But he took one look at the apartment on Exposition Boulevard and he did not give a good goddam if it made sense or not. His room at the Hotel Rue was a rat's nest, and the whole hotel was filled with aging drunks and young, hard-core, mainlining junkies, and a steady stream of whores and pimps roamed the halls that smelled of piss and vomit. He got his things together out of the room, all of which he could carry in one cheap suitcase, went downstairs, and got into Charity's Cadillac where she waited for him across the street.

"That didn't take long," she said.

"You travel as light as I do, nothing takes very long."

He wanted to talk more to her about what she was doing and why she was doing it, but he had already cultivated the habit of not thinking too long or too hard about anything. Slide with it, and don't look back. All he knew was that he was in a great car with a beautiful girl on the way to an immaculate apartment, an apartment such as he never once in his life thought to live in.

He kept his job at Jacques' and was extraordinarily happy in his new apartment. Charity came and went. Sometimes she spent the night, most times she did not. She said she had much work to do at the university and many responsibilities, which did not bother Eugene at all. She began to keep some of her clothes at the apartment and gradually books began to appear, and more books, then the word processor. Eugene had never seen a word processor and he marveled at the way she used it, her fingers flying and then the pages ripping out of the thing full of her words, all beautifully and neatly printed. He could and often did sit for hours watching her work, she occasionally glancing up and smiling at him. How he longed for an education, to read and understand the thick books with raised titles she kept poring over.

By God, whatever else Charity might be, she was a worker. Many times when he got home at three-thirty in the morning she was still at the table under the lamp, an open book in front of her, and Eugene was always amazed at how fresh and relaxed she looked, her face unlined, her eyes clear, her hair—every strand of it—carefully in place, and her clothes as fresh as if she had just put them on, which in most cases she had.

It was several days before he suddenly realized how often she changed clothes, whole outfits sometimes three or four times a day. At one point he started to ask her about all the clothes-changing and the constant trips to the bath to stand under a steaming shower for long periods of time before toweling down and rubbing her body slowly with sweet-smelling oils. But he decided against it. What business was it of his? She was a woman and women liked to look good and smell even better. His daddy had taught him that, saying a man's natural way was to stink, or as his daddy put it, "A man's got the right to stink." But not women. "You take care, son, of any women that don't smell just right. They ain't nothing but trouble. Remember you daddy told you so."

During those first two weeks on Exposition Boulevard she talked to him constantly about his life. What the place and the family he came from were like, what it was like to train constantly, running, sparring, working on the heavy bag, the speed bag. How it felt just before a fight. Did he have to work himself into a rage against his opponent? Did he hate his opponent after a fight? Did he ever hate himself?

"Hate myself?"

"Well, you are hurting people, hurting them physically and deliberately."

"It's a sport, Charity. Hate's got nothing to do with it."

"But it's the only sport where the only object is to inflict pain on your opponent."

"I never really thought about that. For me it was a way out."

"A way out of what?"

"Out of being a nobody."

"Every human being is a somebody, Eugene."

"Easy enough to say if you've got a Cadillac car, you folks ain't hungry and sick, and you with no money to help them."

"That's silly."

"To you maybe. Not to me."

"Tell me about your manager."

"Tell you what about him?"

"How it was between the two of you. Did he almost seem like a father to you? Would you have done anything for him?"

69

"Yes to both questions, but I don't want to talk about it."

"Why aren't you with him now?"

"I couldn't win anymore. A manager needs a winning fighter."

"A father doesn't need a winning son."

"I don't know what you're talking about and you don't either. I ain't talking about my manager."

"There's something very painful there somewhere. I can hear it in your voice. I am right, it is pain, isn't it?"

"It's no fun being a loser after you a winner. And I told you, I ain't talking about that. You bring it up again and I'm gone."

"Sure, let's drop it . . . for now."

"Forever."

"Whatever you say."

And she was as good as her word, it didn't come up again. But she was full of surprises and Eugene did not particularly like surprises. Three days later she came into the apartment with an expensive clothes bag embossed with the name of a store Eugene did not recognize. Eugene said nothing, he just sat watching. But he had the feeling that whatever this was, he was not going to like it. She had a box under her arm and she set it on the bed beside the clothes bag.

"I bought you a little something," she said.

"I don't need a little something," he said.

"Wait'll you see this."

She unzipped the bag and took out a beautifully cut charcoal-speck sport coat, and a pair of light-gray trousers. As surprised and amazed as he was, Eugene still said nothing, only watched quietly. She opened the box and took out a pair of boots.

"Those are Tony Lama snakeskin boots," she said.

"Never heard of them and I only wear motorcycle boots."

"You may be the only man in the civilized world who has never heard of Tony Lama boots."

"May be," he said. "You mind telling me what all this is about?"

"They're clothes for you."

"Why would you want to trick me out like a pimp?"

"Eugene, these are excellent clothes and all the best people in New Orleans dress this way."

"I'm from Bacon County, Georgia," he said.

"Man-child, you exasperate me."

"That's another thing, you got to quit with the man-child bullshit."

"What's wrong with that, honey?"

"My mama named me Eugene."

"My God, my God," she said. "Very well, let's get to the point. I bought these things because I thought they'd please you. We've been invited to a party and I wanted you to look nice because . . . because you're beautiful and you ought to have beautiful things."

"What do you mean party? Why would anybody invite me to a party?"

"Because you know me."

"Not good enough. Besides, I never much cared for parties."

That was a lie. Back in Georgia, he had loved parties. They were given in farmhouses all over Bacon County. There were peanut poppings and candy pullings with pretty girls and guitars and fiddles and the houses smelling of whatever food was available, invariably smelling wonderful no matter the food because it was a party, and—if the weather was right—a big fire outdoors with a couple of pigs and maybe a goat and several dozen quail cooking over a bed of oak coals. So he was lying, but not really lying because this was New Orleans and he knew that whatever a party was in this town, it would be nothing like anything he had ever known in his home country.

"How do you know you wouldn't like it? You've never been

to a party with me. Just try on the clothes, man—Eugene. I went out and bought them for you because I wanted to do it. The least you can do is see how they feel on you, how they fit. That's just common courtesy."

She had talked to him long enough about his family, how he was raised, that she knew the right buttons to push, and she was well aware that she did.

"I don't guess it'd hurt to try the stuff on."

He pulled on the trousers, a perfect fit. He looked at her. "How did you know what size to buy?"

"Darling, you've not just been near me, you've been in me. You can't get much closer than that."

He knew he was blushing, so he turned away to slip into the fine pale blue linen shirt she had bought. It not only fit but tapered nicely to his lean waist.

Jesus, he thought, she must have measured me in my fucking sleep.

"Do the boots next." Her face was radiant. He thought she looked like a child unwrapping presents at Christmas. "Try the boots. There's socks in the box for them."

He pulled the snakeskin boots on over the socks and when he looked down at them, he suddenly felt like somebody else. He didn't quite know who, but he did not feel like Eugene Biggs. They fit perfectly and were as soft as a glove, and he was reminded of the first time he ever saw a brand-new sheepskin headgear, perfectly measured and made to do what it was supposed to do. In spite of himself he felt his heart quicken.

When he turned to her, she was holding the sport coat for him. He slipped into it. Then from a little package he'd not seen, she took out a tie that was just a shade bluer than the shirt he was wearing.

"Now this and we're all finished."

She put the tie in his hands and he simply stood staring at it for a moment and then said: "I've never had a tie on in my life."

"There's a first time, darling, there's always a first time. You might like it."

"I don't think so," he said and stood with the tie hanging from his open palms.

She instantly saw the problem. He didn't know how to put it on, how to tie it. If she was going to get him to the party, the last thing she wanted to do was embarrass him. But she prided herself on never embarrassing anyone unless she meant to do it. As she herself often said: "I'm not getting a doctorate in psychology for nothing."

"Here," she said. "Let me. I just love to knot a tie. At home I always tie Daddy's. I've been doing it since I was a little girl."

She whipped a perfect Windsor knot in the tie and cinched it snugly against his finely muscled neck. "Now," she said, "check yourself out in the mirror and tell me what you think."

There was a full-length dressing mirror beside the wardrobe where she kept her clothes. He approached it from the side slowly and then in a sudden movement like a swimmer diving into water he stepped in front of it. For a long moment he said nothing, and nothing showed in his face.

Then in little more than a whisper, he breathed the words: "Well, I'll be goddam."

"You're quite the man, aren't you," she said.

"I wouldn't believe it was me if I didn't know it for a fact."

"Oh, it's you, Eugene. You'd be beautiful in sackcloth, but now you just happen to be standing in about nine hundred dollars' worth of fine shit."

He could hardly bear it when she talked about him that way. It was embarrassing.

"Why are you doing this?"

"I told you. We're going to a party."

"I never said I was going to a party."

"But you will. My God, I've gone to all this trouble, surely you wouldn't be mean enough to disappoint me." Again, the right button, and she knew it.

73

"Where is this party?"

"At the university. Tulane. Some sorority girls and fraternity guys. You'll have a grand time."

Eugene walked over and did not so much sit on the side of the bed as collapse on it. "You remember me telling you one time I thought you might be just a shade crazy? Well, I don't think it anymore, I know it. What am I gone do at a party with a bunch of college students?"

"You're going to have a good time."

"Charity, you the only college-educated person I've ever known in my whole life. Listen to me, goddammit, and try to hear me. I'm an ex-pug from a dirt farm in South Georgia. That's all. That's it."

"Have you been all right with me? Do you like being with me?"

"Yeah. I feel funny about it sometimes, but yeah."

"Do you trust me?"

"I guess."

"You guess?"

"That's the best I can do."

"It's good enough. Come with me. Please. I've promised everybody they would meet you."

"The more you talk the worse it gets."

"Then let's not talk about it anymore. Let's just do it. Hell, they're just a bunch of spoiled rich kids. Surely a fighter's not going to be scared off by people like that. You were thirteen and 0. Try to remember who you are."

He got off the bed and went to stand in front of the mirror again. "All right," he said. "I'll go."

Eugene had been very nervous on the way to the party, which was going full blast when they got there. Tulane University was on St. Charles Avenue and the party was within a block of it, but it was not in one of the really monster mansions so common on St. Charles, but rather in a house—a house of only two stories with a gallery running on two sides and across the front. New Orleans jazz ripped across the green lawn to meet them as they got out of the car. Laughter and shouts and a babble of voices came with the music. It sounded to Eugene like nothing so much as the monkey house at a zoo.

"Christ," he said, "I wished I hadn't done this."

"Relax."

"It's the first time I ever wanted a drink in my life."

"Then have one."

"Fighters don't drink," he said.

She laughed. "You're impossible. Be easy, for God's sake."

Eugene got a good look at the party through the open door before he even went inside. Actually it looked pretty damn good. Just a lot of young people, his age more or less, some of them about half in the bag from drinking, and on a raised stage at one end of the large room four aged black men playing the jazz he had heard coming across the lawn. He felt his chest lift a little. Hell, this might be fun. God knows he'd had little enough good things in this life since that dreadful bout in Madison Square Garden.

Charity went in ahead of him and he came through the door smiling, feeling that this might not have been such a bad idea after all. But he had taken no more than five steps into the room before it seemed everybody was staring directly at him. Even the music seemed to drop a notch in volume, and people who had been moving about, talking or dancing, grew strangely quiet and still. And then as if on signal, the entire room rushed toward him, voices giggling, talking, squealing now, and he was suddenly surrounded. None of them came closer than ten or fifteen feet but still their eyes were on him and he was the center of the circle. He was baffled and more than a little surprised. He looked around for Charity and could see her nowhere.

A very pretty girl in a short pink dress and wearing some sort of beaded headband came out of the crowd and did not stop until her breasts almost brushed Eugene's chest. Her eyes were very bright and she smelled of liquor.

"Can you do it?" she said.

"What?" said Eugene, not quite believing what he had heard and in any event not knowing what to make of it.

"Can you really do it?"

He didn't know whether to be angered or flattered. "I could the last time I checked."

"Well, come on, champ, let's do it!"

Eugene did not understand anything of what was going on but he didn't like it. He could feel the skin on his face tighten with anger.

"I don't usually fuck in public."

Everyone roared with laughter, even the pretty girl in front of him. "You know I don't mean that," she said.

And that's when Eugene saw Charity. She was standing half-way up a winding staircase looking down on it all, her face calm, disinterested. She had her little notebook in one hand and her gold pen in the other. Then Eugene understood. He understood it all.

"We want to see you knock yourself out," the pretty girl giggled.

Eugene didn't set himself for it or even think about it. The shot that he caught himself on the chin with came out of a rage of hopelessness and despair deeper than blood or bone.

He came to in the Cadillac, the top down, driving slowly through the hot, humid night on a street that had a canopy of huge oak limbs. He had his eyes open, but his senses had not entirely returned. He couldn't at first remember what had happened or where it happened, and he turned to look at Charity, who was staring straight ahead. A kind of shy smile showed on her lips as they passed under the street lamps and her shining hair caught the light. She was humming some tune a little breathlessly. He couldn't make out what it was. But it was some jazz something or other he seemed to have heard recently. And with the tune he remembered it all, all of it in a kind of slow-motion montage of the sort he'd seen in movies. Shame broke on him like a sweat.

"You lousy cunt," he said. He had meant to speak it loudly, angrily, but the words came out slightly slurred as though he were drunk.

She turned to look at him briefly. "Be still, darling, you'll be better in a bit."

"You scum-sucking dog," he said, his voice better now, though still weak. "Trick me out like a pimp and show me off like a clown."

"It was only an experiment. You'll come to understand. It's science. There's nothing personal about it."

He wanted to tell her that he was not a goddam guinea pig, but his head throbbed, his tongue seemed swollen, and his right hand felt like it was broken. He thought he had probably never hit another fighter as hard as he hit himself tonight. He gathered moisture in his mouth so he could speak.

"How did I get in the car?"

"The boys at the party carried you out to it."

He stiffened and sat up on the seat where he had been half-slumped.

"You bitch. You're asking to get hurt bad. You let those puffballs put their hands on me when I was out? I hope I don't kill you."

"I hope so, too. For now, though, we're going for a nice slow ride out River Road until your head clears and you're thinking right again and then we're going back to the apartment and I'm staying with you tonight. Maybe after I talk with you, you'll understand, and maybe you won't. But I don't think you'll leave me and what I'm doing. I think we can strike a deal."

"You think I'm a fucking clown is what you think."

"You'll come to see you're wrong and that I need you almost

desperately for my work. You are unique. Nobody like you will ever come along again."

Eugene was not very good at cursing, having been isolated on the farm growing up and having never heard his father curse in his life, and he was struggling mightily to think of the worst thing he could say to her when she reached across to him and said: "Here," and put a roll of bills in his hands. "It's five hundred and twenty dollars."

"I don't want your money," he said, and tried to give it back.

She kept both hands on the wheel and looked straight ahead. "It's not my money. It belongs to you. It happened just as I knew it would."

"What?"

"That the world smells of shit." It was the first bitterness he had ever heard in her voice, and as angry as he was, he was mildly surprised.

"That doesn't tell me anything I didn't already know."

"The money came from the people at the party. They wanted me to give it to you for . . . for your performance."

"You mean those kids back there laid out five hundred and twenty dollars?"

"They don't call Tulane Jewlane for nothing. They reacted just as I knew they would."

"Jewlane? I don't . . . Oh, yeah, right. That's a fairly shitty thing to say."

"I didn't make it up," she said. "I'm a scientist. I don't invent the world. I only try to discover and describe it."

He held the money in his hand and stared at it for a long time while the Cadillac rolled smoothly out River Road under a full moon.

And that was the way it all started, the way he had come to quit his job at Jacques', the way he had come to be able to help his folks over some hard times back home in Georgia, and finally the way he had been introduced to what he had come to think of as the stinking, hairy underbelly of New Orleans.

The herd of mongoloid children came galloping past the bench again where Eugene sat. He marveled at their stamina. Their keepers were huffing and puffing but the children were laughing and squealing and bucking along like young colts. A desire seized Eugene to get up and run along with them, to laugh and forget what the world demanded of him, but he had kept Charity waiting too long as it was.

When he came into the air-conditioned apartment she was lying on the bed, naked and on her back. The curling golden hair at the base of her belly glowed like a light. The fragrance of sweet body oil came to him all the way across the room. A few drops of water still clung to the oil between her breasts. The black Dictaphone that he had come to hate sat on a small table beside her bed near the satin pillow on which her head rested, her hair fanning out from her shoulders in perfect symmetry like some beautiful growing thing found on the bottom of the sea.

"I want it all," she said. "All of it." Her voice was soft and flat, entirely without inflection.

He took off his clothes and lay beside her. Her skin was very hot, whether from standing under a steaming shower or from the passion of this ritual of theirs, he did not know. But he was accustomed enough to it now not to be disturbed by it.

She drew him to her and kissed him and he felt his pulse jump and the quick fire leap in his loins in spite of what he knew he was obligated to do.

But this was the deal he had struck. If this was it, this was it. You give up some things in this world to get some other things. Luck of the draw, maybe. Who the hell knew? He only had to do it until he decided not to do it.

She was still hotter and wetter there where he joined them together in one savage plunge and she sighed deeply, "Yes," and in the same instant he heard the soft whir of the Dictaphone start up.

He told her about counting the suits waiting to go out and

do the gig the night before and about entering the huge room where he encountered the light heavyweight being harangued by her diminutive manager.

"And she was naked from the waist up?"

"Yes."

"And others?"

"Others, too."

And he told her about the caricature of a ring he entered and the ridiculous spit bucket and about Jake.

"And Jake was beautiful and a lesbian?"

"She was beautiful."

"And a lesbian."

"I don't know."

He told it all, the Dictaphone whirring while the two of them tossed over the bed in a kind of battle, she sometimes in a four-point stance and he riding her like a college wrestler, and other times he being ridden, the sweat pouring off her chin onto his face, the bedclothes soaked now, and still the telling went on and the whirring of the Dictaphone.

"But what was his real name?"

"I don't know."

"Oyster Boy? Why Oyster Boy?"

"I don't know."

"Didn't you want to know?"

"No."

"You wanted to know. Admit it. There's no need to lie here."

"I wanted to know."

"You could have found out."

"I . . . I . . . No, goddammit, NO!"

"And he was on a leash."

"Like a dog."

"And scaly?"

"His skin flaked when he scratched."

And he told her about going to the Quarter and talking to Pete and about the snuff film.

"And you watched it?"

"I glanced at it once. Then sat. Where I . . . couldn't see it."

He was under her now and she ground her hips mercilessly and he felt his brain would burst.

"But you wanted to stare at it until your eyes cracked."

"Yes. Yes. Yes."

And he told her about Pete and Tulip and how Pete had been with her and of the act she had with the teddy bear and how the act worked.

"She hits a switch and a cock springs out of the teddy bear?"

"Like a goddam switchblade knife."

He was going now. It was like being washed down a roaring river. And he could see the falls ahead. He was going over the falls and they were rushing toward him with the water roaring in his ears.

"And she pretends to fuck the teddy bear?"

"Not . . . pretend . . . No."

He could see the abyss beyond the falls and the drop was long, very long, and he thought his brain would burst.

"She fucks the teddy bear?"

"Yeeessss."

And he burned all along his nerves as he dropped into a place where the world did not exist, all of it entirely destroyed.

They both lay gasping, clinging together like drowning swimmers. He felt her hand move across his chest, her arm extend, and the whirring Dictaphone went silent as she touched it.

Chapter Five

Eugene was sitting at a little rolltop desk that Charity had bought for him because she said he needed his own place to work, just as though he read and wrote and sat thinking as she did. It had pleased him at the time she bought it, but the only thing he did there was to write infrequent letters to his daddy, which he was trying to do now while Charity sat on the other side of the room wearing a headset from the Dictaphone, her fingers going ninety words a minute, transcribing what he had told her the night before.

It was a strange and distasteful way to have a conversation or to be questioned about the most intimate aspects of his life, but it was part of Charity's own private theory of how best to get at what she called deep truth. As she had explained it to him, everybody's defenses are totally destroyed during a raging fuck. He didn't really know what she was talking about and tried not to spend any time thinking about it. It was a deal he had made. And his daddy had raised him to believe that a man's word was his bond. A deal was a deal.

He read his daddy's letter again, and then wrote quickly on his notebook pad:

I was sorry to hear about them calves, but I know mama and Edsel will get better because they are strange

He looked at the word *strange* for a moment and then wrote *stronge*.

and hear is a thousand dollars. I won again by a knockout.

your son,
Eugene.

That was all he could bring himself to write. He wished he could write a longer letter, but all he could think of to write were lies. And lies were beginning to sicken him. He put the letter and the money in an envelope, but as he started to address it, he heard the word processor and whirring Dictaphone go silent, and when he looked up, Charity was watching him.

"I've got to meet him, Eugene. I've got to."

"Why don't you listen? I don't even know his name. And I don't want to know it."

"And Jake, too," she said.

"That can't even be her name. That was all a game them people were playing that I don't understand a thing about. I got no notion at all how to find her."

"But you could find him. And as far as understanding, I don't, either. But I want to. It's my life."

"You've got a shitty life."

"I don't think so."

"You ought to do something about the way you live and the things you think about."

"I won't be unkind enough to tell you that you are obviously not the one to be making such a value judgment."

"And about the way you talk, too. You talk funny. You talk like that where I come from, they'd hang you."

"I'm not where you come from, darling. And my purpose in life is not to live up to your expectations."

"Your fucking what is not what?"

"I can't do things as you want me to do them. I have to do them according to my own best lights. If they are wrong, I'll just have to live with them."

"Or die with them."

"That, too."

"You're not going to use my name in that stuff you're writing, are you?"

"How many times do I have to tell you? No. Your name is of no consequence. The source of everything I get is confidential, secret, left out. O.K.? And you could find Oyster Boy and arrange for me to meet him. I know you could."

"It was bad enough once. I don't think I'll go for twice."

"I'm not asking that of you. I'm asking you to help me. I thought that was the understanding we had."

"I've got to address this letter and get it in the mail."

"Sending money home, are you?"

He did not look up but kept on writing on the envelope.

"Are you going to make me remind you that if it weren't for me you wouldn't be sending that home to your parents?"

"You just did whether I made you or not."

He heard the Dictaphone click on, but only for a moment, before it went off again.

"Tulip sounds fascinating. Maybe you could take me to see her perform sometime."

"You want to see her, it's a five-dollar cover at the door. Five bucks. You don't need anything else. Three-drink minimum at four dollars a shot and the whiskey's watered so you might as well not drink it."

"I'm afraid of those places by myself."

"You're safer in there than you are on the street. There's guys in there to make sure you're safe."

Charity hit the button on the printer and all that she had put into the word processor from the Dictaphone came ripping out onto page after page. Eugene, watching, was acutely aware that those were his words, the things he had said, that were being put down in line after line of black type.

She finally turned off the printer, took the pages and the tape out of the Dictaphone, and bent to the heavy padlock on the steel filing cabinet with a key.

"What all exactly have you got in there?" he had once asked her.

"I've got notes," she said, "and a life in there."

"A life?"

"A life, a book . . . same thing. A book is always a life. Sometimes the life of the one who wrote it, sometimes the life of someone else. Sometimes both."

"That shit's too thick to stir with a stick," he had said and walked away. He never asked her again about what was in the cabinet. But that did not keep him from wondering about it. Sometimes he caught himself staring at it, thinking about ways to break it open without her knowing.

Charity turned from the cabinet and said: "If Pete's seeing Tulip, he could bring her to dinner."

He could only look at her and shake his head. "Charity, to have been in school so long you don't know your ass from a hole in the ground."

"Quaint," she said, "a really quaint saying. You're trying to piss me off. Well, it won't work. Science does not anger."

"I wish you would stop with the science. I'm tired of hearing it. Listen, you want to meet Tulip because you think she's . . . that she's got some kind of secret. That you're going to be let in on some kind of magic . . . that she knows something you don't know. Let me tell you how it is. You meet her, right? What do you find out? Something like she's from Oklahoma and the job she had before this one was selling perfume in a department store. And if she was old enough—which she ain't—you might even discover something really ugly and horrible, since it seems like that's what you looking for anyway, something scary and horrible like after she gets through doing the number with the teddy bear she goes home to her husband who's a carpenter out of work, and she loves and cooks oatmeal for her seven-year-old daughter before she sends her off to school." He could feel his face flushing with anger. "You've got your head up your ass, girl."

"Maybe, maybe not. She may be that or something else. Either way it would not surprise me."

"So you just like to peep on people, get your jollies peeping on people, is that it?"

Her beautiful mouth went thin and her face hardened. She stood up, her shoulders trembling as though from a chill. "No, you goddam ignorant bastard, that is not it. Why can't you let me get on with my work without this constant interfering bullshit?"

"You know who half the people were going into the snuff film last night?" he said. "Fucking teachers. Wearing those fancy little shirts saying 'Teachers Are Precious.' "

"I know, dipshit. I was there."

She stormed out of the room, her sandaled feet hitting the floor much harder than they needed to, and a moment later he heard the shower running.

He sat listening to the shower. So Charity was at the snuff film. She gives some kind of thing in front of a teachers' convention and then goes watch a guy get butchered alive. He could see her sitting there in the dark, not hooting and hollering and groaning with the rest of the audience, but with her little notepad on her knee and writing in the dim light from the screen with her gold pen. So what? It was none of his business what she did. He had enough to think about without worrying about her problems.

He did what he often did to calm himself, to take his mind to another place. He opened one of the drawers on the little desk and took out some pictures, some of them eight-by-ten glossies and some of them much smaller, dim and yellowed. They had been given to him by Budd, and Eugene had taken them with him everywhere he went. He had them in his suit-case when he came to New Orleans for that last fight with Pete. Budd had left in such a hurry that he had not had a chance to give them back. He could have sent them to him in the mail and he had started to more than once, but he simply could not bear to part with them.

The pictures were all of fighters. The one on the top was of Jake LaMotta posed in a fighting stance in the ring with his gloves still on just after he had destroyed some poor bastard. Eugene had seen the movie of LaMotta's life called *Raging Bull.* Raging Bull, my ass, thought Eugene. La Motta had been known as the Bronx Bull. Well, Hollywood, Hollywood would fuck up a wet dream. Everything had to be changed around and twisted and screwed up until it became something it was not.

He felt the heat in his face and realized that he was not thinking about LaMotta, he was doing what he had carefully cultivated the habit of not doing. He was thinking about himself, about his own predicament, and about Charity and finally—startled—he realized that he was thinking about Jake. He could not remember thinking about her once since the performance at the mansion on St. Charles until he told Charity about her, and even then he was not thinking, he was just talking.

He quickly shuffled through the pictures. The beautiful, unmarked, slick-haired Sugar Ray Robinson, nattily dressed and surrounded by fine women in a club somewhere on the very night when he had won the world for the third time. And then he sifted through a series of action photos of the great fights Billy Conn had given Joe Louis. The last of the series was one of Conn flat on his back with Louis turning to walk to a neutral corner.

He smelled the sweet body oil before he felt Charity's hands on his shoulders. She massaged them, squeezing and kneading in a steady rhythm.

"Can you forgive me?" she said. Her hair brushed damply against the back of his neck as she kissed the top of his head.

"You don't need me to forgive you. Ain't nothing to forgive."

"I'm sorry anyway. Let's not argue."

"We weren't arguing. Forget it. It don't mean nothing."

"You were looking at your pictures," she said. "It's really quite some collection."

She had said much the same thing before, more than once. He did not say anything.

"And they didn't want it any more than you did, did they?"

"Nobody knows. Caint know that."

"But you wanted it more than you ever wanted anything before . . . or since, didn't you?"

"I don't think about it much."

She patted his chest with both hands. "That's all right, darling. We'll talk about it another time when we're . . . closer."

The telephone rang and she walked to the table beside the bed to answer it. Eugene never got a call at the apartment except from Pete now and then.

"It's for you," she said.

"Christ, B and B probably broke his face."

"It's not Pete," she said.

He took the telephone. Charity sat on the side of the bed watching him. After saying "Yeah," Eugene simply stood listening, nothing showing in his face at all. Then his eyebrows jerked and his mouth tightened. "Two thousand? That's like one, two, two you say? And nobody's fucking watching? Oh, yeah. Fucking. Watching. Well, what the hell. What do I think of it? I don't spend a lot of time on shit like that." He picked up a pen lying beside a pad on the table. "Give me the numbers, man. What do you think I mean? Where? When?" He wrote on the pad. "Before you go, one thing. The money at the door or it's goodbye, Charlie."

When he hung up the phone, Charity asked: "Who was that?"

"I don't know."

"But a job, right?"

"Yeah, a job."

"Tell me."

"You'll know when I know. That's the deal. I don't want to talk about it."

"I heard you say two thousand. Was that dollars?"

"That's what the man said."

"Better and better," she said. "My God, they may want you to open a vein for that kind of money."

"I'll deal with it when I get to it," he said.

But he was as shocked as she was. The fifteen hundred he had made at Oyster Boy's was the most he had ever made. But there had been three hundred people there, maybe even more, and it was in a mansion on the most expensive avenue in New Orleans. So even though it was a record payment, it didn't seem entirely out of line. What he had just heard over the phone seemed not only out of line but more than a little crazy. He thought about the long-barreled .38 revolver and the shoulder holster in the bottom drawer of his little desk. He would have to go armed. He did not want to, but he thought he had better. There was something about this that did not smell right. But at the same time, wearing a weapon was the surest way to wind up killing somebody and being locked down for the rest of your life. Better that, though, than fall into some kind of sick shit where he not only died, but died at the slow pleasure of a pain freak with a lot of patience and in no hurry. He thought of the snuff film.

The address he had been given was for a condominium on Canal Street, which—this being New Orleans, where nothing was what it seemed—had no canal on it or near it. It was a broad thoroughfare that formed the western boundary of the French Quarter. The man who talked to him gave him no name, and Eugene had asked for none. That was only to be expected. That he was to come alone or the deal was off did not surprise him either, but the rest of it did.

He would be met at the door by a houseboy and taken to a bedroom where a man and a woman would be doing a straight number. He was to stand at the foot of the bed and perform on command. And there was one final kicker: Nobody would watch. What the hell, he didn't need an audience. But his daddy needed two thousand dollars. If it worked like the man on the phone said it was going to work, it would be found money. Half the farms in Georgia had already gone under from the deep depression that had struck agriculture, but his daddy's farm, as sorry as it was, was going to remain in the family.

He even dreamed of getting lucky somehow and buying his daddy and Edsel a really good farm, not a big one, just one with a tight modern house on it and good, rich bottomland. If that ever happened, he would be right back there with them. He would simply tell them that he had run out of gas, couldn't crack the nut anymore, and had retired. They would be happy about that, especially his mama. She worried constantly about him fighting. It was one of the reasons he rarely went home. The time he went home with scar tissue in his eyebrows and a recently broken nose, the nose broken not in a fight but sparring in the gym, she had collapsed when she saw him. She didn't faint; all her bones just seemed to turn loose and she slid to the floor. His daddy had said in his letter that she was in bed

90

now with female trouble, whatever the hell that was, but he had wondered from the moment he read it if she might not in fact be in bed because of nervous days and sleepless nights worrying about him getting hurt in the ring.

But what would he tell them he was retiring for when he had written them all those letters about a string of knockouts such as no fighter had ever put together? Nothing, of course, because they would not ask or care. None of his family had ever seen a fight, not even on television, or so they had told him and he believed them. He had tried to get his daddy to let him buy them a television set, but his daddy had said it was a waste of money and besides there were too many other things he needed. Eugene had never known his daddy or his mama to go to a movie. Edsel had been to the one in Alma and the one in Waycross, but he said he didn't care much for them either. He would rather fish or hunt. At least that put meat on the table.

Eugene sat at his little desk slipping rounds into the chamber of the .38 before replacing it in the holster in the drawer. He could hear Charity in the kitchen as she moved about making an early dinner. He could count on the fingers of one hand the times he had ever seen her cook, but she was surprisingly good at it, excellent, in fact.

"I'm good at a lot of things I don't do," she had told him. "For me to cook would be like washing and ironing my own clothes. It would be a waste of time, and time, sweetheart, is all we've got."

But when he had told her that it was on for midnight tonight, she had insisted on cooking his dinner and eating with him.

"I'll have to go right after dinner, but you just relax until it's all ready."

He could not relax. Invariably, waiting was the worst part of it, thinking about what he was going to do and why he was doing it. When he walked into whatever he was walking into tonight, his daddy and mama would have been asleep for

91

hours. They hated to turn on the lights. They were not stingy; they were only squeezed as tight as they could get for money. If they ever found out about the things he had seen, the things he had done, the places he had been, their hearts would stop on the spot.

He knew that fantasizing about buying a good farm and going back to Bacon County to work it with his brother and his daddy was just so much bullshit. He would buy it for them in a minute if he could, but he could never live on it, not now. He had become somebody else during the last two years. What had happened to him was irreversible. It was almost unbearable to think about that, and most of the time he could avoid it. But tonight, perhaps because of the letter he had written during the day, the thinking he usually could shake off was heavy on him.

He went over and lay down on the bed and closed his eyes. And even though he knew something in him had been permanently scarred just as his eyebrows had been, scarred beyond any hope of ever fitting back in with the place and people he had left at seventeen to go to Jacksonville, even though he knew that, behind his closed eyes he saw himself back on the land in Georgia, back where people talked straight and where things were what they seemed, where he could see unfolding plants break ground in spring and grow green and lush until harvest, where he could meet a young girl who was neither a whore nor somebody who needed a Dictaphone to take him to her, and where all his friends would have ordinary jobs. Ordinary. That was it. That was what he, at one and the same time, was heartsick for and the very thing that he would never again be satisfied with. He sat up on the side of the bed and rubbed his face.

Very softly he said to himself, "Well, you brainless bastard, are you happy now? See all the good you can do youself by thinking, chewing on your life and the life you ain't got like a cow chewing on her cud. Is that enough? Or do you want to lie back down, close your fucking eyes, and bleed some more?"

"Eugene," called Charity. "Let's eat it."

He got up, not wanting much to eat, but changed his mind as soon as he stepped into the kitchen.

"Where did you get that stuff?"

"A place I found because of you. I knew there would be times when I'd want to do this for you."

She had done it only once before, cooked the food he grew up eating but rarely found in restaurants. He did not remember telling her the kinds of things he liked to eat, the sort of stuff his mama cooked. But obviously he had. In their raging sessions in bed, he knew he had to have told her. He sometimes thought she knew more about him than he knew about himself.

On the table were a beef tongue surrounded by new potatoes and chopped green onions, and a platter mounded with golden tripe, and plates of tomatoes and onions, and hoecake and corn bread, and beside his plate a tall glass of buttermilk.

He stood shaking his head. "I don't know. Every time I think I got a handle on you, you do something like this. Why did you go to all this trouble? You don't even like this stuff."

"Are we going to stand here and talk, or are we going to eat it?" She pulled out a chair and sat down. "And don't tell me what I like and don't like, because that, dearly beloved, is something of which you have no knowledge."

It was true that the other time she had cooked not these dishes, but food of this kind, she had eaten long and with great relish.

Eugene was forking tripe onto his plate and slicing the steaming tongue. "A man could cut pulpwood on this kind of food."

"Fortunately we don't have to do that. But I do have to go when we're through here."

"They can take all the sauces, French or whatever, and

dump'm in the shitter for all I care. I wish I could eat this meal every day."

Charity looked over the glass of red wine she had poured for herself. She said buttermilk was rather much to ask of a civilized person because it tasted as though it had not only soured but perhaps rotted as well.

"If you want to eat this way, there's a stove in there and I'll be happy to show you the shop."

"I can't cook."

"I know that's what you say. But you could learn. It can't be difficult or so many people would not be doing it."

"No man in Bacon County can cook. Maybe an egg or something. But cook? None of them."

"You could learn. I don't like it when you keep saying you can't do this and you can't do that."

"Try not to think about it."

"What are your career goals? Do you have any?"

"My what?"

"What do you plan to do with yourself, your life?"

"Where the hell do you get these questions? We were talking about cooking."

"We were talking about learning. I want to know what you intend to learn, intend to do."

"I haven't made any plans."

"Did you ever think you might become a criminal?"

"You trying to ruin this for me? What kind of thing is that to ask?"

"Now come on, we're only talking here. We can have a conversation, can't we? You are constantly around things that are illegal. You said yourself that there was every imaginable kind of drug where you were last night. I care about you, silly boy."

"I'm not silly and I'm not a boy, and no, I've never given much thought to being a criminal."

"That's the point, you're a man, and a hell of a man at that. Do you ever give much thought to anything?"

He put down his knife and fork, and looked at the ceiling as though the answer might be written there. "As a matter of fact, I do. And you ask me another question, I'm gone."

"No more questions, boss man."

"I'm nobody's boss man but my own. And I even eventually manage to fuck that up pretty good."

"We're all too hard on ourselves. For God's sake, try to enjoy the food. I don't do this kind of thing very often for anybody, including myself."

"I can't imagine where you learned to cook like this."

"Books, Eugene, books. I love this sort of food every now and then and given the time, I prefer to prepare it myself."

"I never learned a goddam thing from a book."

"I'm not going to touch that remark."

"Good," he said.

She pushed back from the table. "I wish I could stay, but I've got to run. You'll have to put the dishes in the washer and put what you want to keep in the fridge. That tongue will make great sandwiches."

"No problem," he said, taking more tripe and tomatoes onto his plate. "It was good of you to do all this. I appreciate it."

She came around the table and put her hand against his cheek. "You be careful out there tonight when you go among the wolves."

It was something she had said before. He only smiled at her, nodded his head, and kept on eating. She went back through the apartment to pick up her purse and he heard her let herself out the front door. Strangely, when he heard the door close, his appetite faded. He moved the tripe around on his plate for a moment, telling himself it was a waste of good food not to finish it, thinking again of his folks. Tripe was one of his daddy's favorites, and consequently Eugene had grown up eating it, and it had become one of the things he had come to like best, too. He pushed his plate away and finished the glass of buttermilk.

He got up from the table, meaning to clear it, but he only looked at it for a moment and said to hell with it, let it rot. He was already getting his game face on for later tonight. He went in and took the holstered .38 out of the drawer and set it on the desk top. He supposed he would have to tape his hand ahead of time. When he had used his bare fist that night at the Tulane students' party, he jammed the knuckle of his forefinger badly. It was a wonder he had not broken a bone. There would be no gloves tonight, but that would only make it easier. He seemed to be set, but he had time to kill. It was times like these when he often went to the Audubon Park Zoo, touted as the best in the nation. If anything, it would finish getting his game face on and put him where he needed to be, the zoo would.

It was right across Magazine Street from Audubon Park, so close to his apartment on Exposition Boulevard that he could hear the lions roar at feeding time and sometimes in the middle of the night from his front porch.

When he got to the entrance and handed the lady his five dollars for a ticket, she said: "We'll be closing in only thirty minutes, sir."

"Just right," he said. "That's all I need."

Inside, despite the time, the place was still full of people, many of them wearing hats or T-shirts or both bearing the legend: TEACHERS ARE PRECIOUS. Eugene thought half the teachers in America must be in New Orleans. The zoo was enormous, filled with lagoons, tropical plants and flowers that were ablaze with color, waterfalls, and more than one thousand animals, none of them in traditional cages, but caged nonetheless. Eugene had spent many hours here, usually when he was feeling a little desperate, or filled with free-floating anger or anxiety. The zoo gave him the same dreadful pleasure that he got from touching an extremely sensitive and aching tooth with his tongue. The pain could be enough to make it seem his whole mouth would explode, but his tongue would keep finding the tooth and probing it anyway. Eugene had seen everything in the zoo many times, the snow-white tiger and twin baby orangutans, white rhinos, clouded leopards, Asian elephants, tapirs, wolves, cheetahs, as well as the animals native to Louisiana, these last contained in something called the Louisiana Swamp Exhibit, the huge alligators, black bears, and all manner of snakes.

But today, as he often did, he went directly to the lions. They were kept in a deep pit maybe thirty yards square, with a stream running through it, along the shores of which young shade trees grew. The pit was enclosed by sheer-faced rock. Around three sides of the pit was a railing made of metal pipe at which the tourists could stand, leaning on their elbows and watching to their hearts' content. Eugene found a place at the railing—not without some difficulty—but when he looked down the lions were not there. A cave had been cut in the face of the rock beyond the stream and little shade trees. That's where the lions were, and the tourists had been waiting for God knows how long for them to come out.

As if on signal, when Eugene touched the railing a heavily maned, magnificent lion came ambling out of the cave, his huge head moving from side to side. He stopped and stretched, his long, lean muscles rippling in the sun. The tourists groaned much as he had heard them groan at the snuff film. This was the way all the animals were kept, in their natural habitat, as the zoo brochure pointed out.

A blue-haired lady wearing a TEACHERS ARE PRECIOUS T-shirt said: "Isn't it just wonderful that they are not kept in cages?"

A slender young man standing with her, wearing not a T-shirt but a straw hat advertising the same preciousness of teachers said, "The marvelous king of beasts. It is indeed extraordinary that they have constructed such surroundings as to make him feel at home, to feel as though he is truly still in the jungle."

The lion looked up for a long moment at the tourists hanging over the railing, yawned, then turned his backside to them and took a lingering, luxurious shit.

There was nervous tittering from the tourists and a child said: "Look, Mommy, he's making do-do."

"Hush, Vernon," his mommy said.

Eugene looked down the railing. That's right, Vernon, shut the fuck up. He would have dearly loved to take Vernon by the seat of the pants and the nape of the neck and throw him to the lion, Vernon along with the slender, precious young man who thought the caged lion felt as though he was truly still in the jungle. Yeah, Eugene thought, the lion was down there saying to himself: Well, here I am still in the old jungle. Wonder what those things hanging up there staring down at me are? Whatever they are, they're too ugly to eat.

Eugene had had enough and he walked quickly to the exit. Back at the apartment he strapped on the shoulder holster and slipped into a light leather jacket he used for riding in the summer. It was constructed to allow air to flow through it from disguised openings along some of the seams. It was light enough to wear in hot weather, but would save a rider a lot of skin if he dropped his bike at speed. It might be a little uncomfortable, but it was best to wear it if he was going to carry a weapon.

He went through the kitchen, where the tongue and tripe were congealing in a gauze of fat, and out the door. He needed to get out of the apartment before it started becoming smaller and smaller. It occurred to him that doing these gigs was getting more and more difficult. He concentrated on the lion at the bottom of a pit from which escape was not only impossible but unimaginable.

He hit the start button on the BMW, cracked the throttle,

popped the clutch, and was off down Magazine Street riding as he never did unless it was just before or just after a performance, weaving through traffic with the throttle open, passing on the left, passing on the right, not particularly caring which. He cut over to St. Charles to Esplanade Avenue where he turned west on Decatur Street, completely circling the French Quarter, roaring all the way, sometimes a little early on a red light, sometimes a little late, and sometimes right through one, weaving, getting on the back wheel at times, laying the bike dangerously low at corners until he got to Jackson Square where he pulled to the curb and parked. The ride had left him damp with sweat, but he felt better.

He thought about going by to talk with Pete for a while but he could not bear to think what might be on the screen, what kinds of sounds might be coming out of the audience, and he sure as hell did not need another dose of whatever may have happened between Tulip and Pete. So he walked over to the Gazebo, an outdoor café where jazz and ragtime piano was played. Sometimes there was a blues or jazz singer and sometimes not, and there was always dancing on the little raised floor under the roof where the musicians played surrounded by an open courtyard. The only people Eugene had ever seen dancing there were old people, the really old, sometimes alone, sometimes in couples, some of them stone sober and some of them half in the bag, because this being New Orleans, liquor—or as the management called them, specialty cocktails—was served there, some drinks pretty ordinary like piña coladas, but some exotic and potent enough to knock your dick in the dirt.

When the young waitress came he ordered a piña colada which he had no intention of drinking, but tables were at a premium and the waitress would only hassle him if he didn't order something. He wished he had not eaten and he probably would not have, and certainly he would not have eaten as much as he had, if Charity had not cooked the food that he never got anymore and always missed. Eating a meal was no way to prepare to go into the tank. It would be all right, though; it would have to be all right. He would go down for two thousand dollars if he had broken ribs.

On the little floor the old people were gliding about with their eyes closed, their faces beatific, slow and graceful, and beautiful, to jazz being laid down by a trombone, a trumpet, a bass, and a piano. There was a fat, sweating white singer with a loosened tie and open shirt, and a huge white handkerchief in his left hand, who sounded exactly like Louis Armstrong and mopped his face in the same way Armstrong used to do. He was so into what he was doing, so lost in it, that it was hard to believe he knew where he was, or maybe even who he was.

Eugene watched one couple that must have been seventy-five, who glided through movements that only a very long time pulling in double harness could have produced. Their lids drooped, their mouths moved along with the fat singer mopping his face as he sang. Occasionally their drooping lids fluttered and they looked at each other, and their dry, cracked lips would brush together.

There it is, thought Eugene, there it is right there. Difficult to find, and once found, nearly impossible to keep. He held out very little hope for such a thing in his own life. He felt that something at the core of him had been twisted and twisted and twisted until it had finally broken. But he was deeply moved and he thought again of his daddy and mama and the life they had had together. The waitress came again and he told her to take away the piña colada. He had decided against it, he said, and told her to bring a double Stolichnaya from the freezer instead. Fuck it, he had already eaten. A couple of drinks wouldn't hurt. If they did anything, they would only get him deeper into what he had to do. Make him angrier. Anger was always a necessary part of it. He never knew toward whom or what the anger was directed, certainly not toward the people who were paying him. They were not at fault. Nobody was making him do anything. But anger was always there nonetheless.

Charity would be bored witless by this kind of thing. She had come here to the Gazebo one time with him and stayed about ten minutes.

"You stay and enjoy yourself," she had said. "I've got more important matters to attend to."

What those matters were he had absolutely no idea. And did not want to know. He had enough to deal with already without having to take on anything else. That was why he had not asked her where she had to go in such a hurry after dinner. He never asked; she never said. For all he knew she had a whole line of people like himself that she ran just as a trapper runs a string of traps. The good side of all that was she never ragged him if he did not show up when he was supposed to show up, did not give a shallow damn about whether he had been with another woman or not. As she had carefully explained to him, jealousy, like marriage, was dead, finished, over, not justified. Well, thank God, he did not care what she believed about marriage or jealousy or anything else.

After four ounces of freezing vodka, off which he caught a gentle buzz, he knew that he had best take his feet and get on to somewhere else. He did not think there was any possibility of screwing up a chance to make two thousand dollars, but until he had the money in his pocket he was not going to tempt himself, get his head wrong or his fist wrong. He walked over to the Farmers Market, a huge place where fresh fruits and vegetables were sold twenty-four hours a day and—as the signs clearly said—which was the nation's oldest continuously operated public produce market. How strange to find a place like this on the edge of the French Quarter. Any good major leaguer could stand in the middle of it and throw a baseball to Bourbon Street where Tulip—where hundreds of Tulips—did their act for drunken, howling crowds of lawyers and doctors and accountants and teachers and every kind of working stiff.

Eugene strolled slowly among the stalls and watched people thumping watermelons and smelling cantaloupes and examining bunches of carrots and radishes so fresh they still had the dirt they were pulled from on them. He had been here many times before because it reminded him of home and how he had been raised to manhood. Most of what he passed and smelled and touched he had at one time or another planted and harvested himself, and his daddy was still doing it. And nearly every time he had been here had been just before he was about to go somewhere to do the only thing left him to do that would make him the kind of money he had to have. All these things from the field, so recently planted and plowed, caused such a keen sense of loss to come into his heart that it drove him dangerously close to the edge.

Finally, when he had had as much as he could stand, he went back to his bike and rode down to the docks that never rested here in the second busiest port in the world. If he walked the docks long enough he knew he would hear most, if not all, of the languages of the world spoken and see every kind of ship that sailed the oceans. It was such a foreign, alien place that it was the perfect place for him now. Walking these docks he could hold his thoughts and emotions absolutely still.

So for the next hour that was what he did, walking slowly, thinking about absolutely nothing, not even his family, watching everything but seeing very little. And at some point, he stopped under a lamp, checked his watch, and saw it was time. He went back to the BMW and started for the address on Canal Street, riding much more calmly now. Everything was settled. His mind was clear and empty. He had done all that he could do to make himself right. He was ready.

The number he had was for a seven-story building, very old, but expensively and apparently recently refurbished. There was ironwork all over it of the sort that was everywhere in the Quarter, twisted wrought iron over the windows and along the balconies, and at the sides of the windows there were the louvered shutters that were on apartments and houses throughout New Orleans. At the entrance, a dark green canopy extended over the sidewalk from the door to the curbstone. On the side of the canopy in foot-high letters were the words: JEAN BAPTISTE LEMOYNE ARMS.

Eugene parked about thirty yards away and, still sitting on the bike, took a roll of tape out of his jacket pocket and patiently and carefully taped his right fist. From time to time as he wrapped the tape, he flexed his fingers to make sure it was tight but not tight enough to hinder the circulation in his hand. There were no guarantees about how these things would go down. He knew from past experience that something could happen and he might end up wearing the tape half the night. When he was satisfied, he stuck his right hand in his jacket pocket and walked to the swinging glass doors and stepped into a lobby where a huge black man in a blue uniform with gold braid and a blue cap said: "May I help you, sir?"

"Eugene Biggs for number five fifty-six. I'm expected."

The black man picked up a telephone and dialed. "There's a Mr. Biggs here in the lobby. He says he's expected." He smiled. "Of course. Right away." He turned to Eugene and pointed. "The elevator is just there. Fifth floor, third door on your right."

The elevator was silent and swift. Eugene raised the brass knocker on the door and let it drop once. It was immediately opened by a thin Chinese boy of perhaps nineteen wearing a yellow silk kimono embroidered with red dragons. "You quite punctual. That good, very good. Come, please."

The young Chinese had rouge on his cheeks and eyeliner over the longest lashes Eugene had ever seen. His voice was high and lilting and pleasant. Eugene thought: A Chinese fag. The whole world is going queer on me. Well, what the hell. Everybody's got to make a living.

"You wait here, please."

"Sure," said Eugene. "But before you go, this is where the money changes hands. You holding?"

The young Chinese blushed, or seemed to, and without a word reached into the folds of his kimono and handed Eugene an envelope. Eugene opened it. Hundred-dollar bills. Twenty of them.

"Take all the time you need," Eugene said.

The young Chinese walked away swaying gently and gracefully and disappeared behind a door. Eugene looked around to see what sort of place he had fallen into this time. Nice, very nice. He could see into the big living room from where he stood. It was painted in dark earthen colors of the sort he had never seen in an apartment before. All the furniture looked old-fashioned to him, but it also seemed new at the same time, highly polished and a deep walnut color, chairs with gold seats and curving, spindly legs. A huge carved desk that looked like it weighed a thousand pounds sat against windows covered with dark drapes that were of a color he did not recognize and had pull cords of dark yellow. He took a step forward and looked deeper into the room and saw at the far end a fireplace bigger than the one his mama and daddy had at home in the farmhouse. A first time for Eugene, a fireplace in an upstairs apartment. He knew it was just for show, though, because it got colder than a bitch in New Orleans in the wintertime and a fireplace could never heat a place like this.

The houseboy came back into the room, smiling and dipping his head. "It time, sir. You follow me, please."

"Whatever you say, pretty."

The houseboy's eyes went brighter and his smile deeper. He took Eugene to a door at the back of the apartment. "To be done here," he said and walked away.

Eugene opened the door and then closed it behind him, deliberately avoiding raising his eyes until the door was shut. When he did look up, what he saw was the naked back of a man on a bed. All he could see of the woman beneath him was her knees, and her widely spaced feet—toenails painted red—and her thin arms and long-fingered hands holding tightly to the man's back.

The man's buttocks were dimpled, his back thick with fat. He was not young, either. It was the back of a man in his middle forties, maybe older. A little web of blue veins traced an intricate pattern on the backs of his hairless calves. But he was working like a young man, driving whoever was under him like a truck.

Then Eugene saw something that startled him, then made him want to laugh, and he never wanted to laugh at a time like this.

I'll be go to hell, he thought, the sucker's got a stocking pulled over his head.

Breathless and muffled from behind the stocking, the man said: "When I say 'Do it now,' do it. Understand?"

"Got you covered, chief," Eugene said. "I understand."

He pulled his taped fist out of his pocket and flexed his fingers. No telling how long this might take. Somebody, presumably the man, since he was the one who spoke, had some heavy cheese tied up in this action. Eugene set himself anyway, and relaxed, his mind utterly without thought.

But it was hardly thirty seconds before the man's soft buttocks started trembling, his back arched, and he screamed, his voice surprisingly clear behind the stocking: "DO IT NOW!"

And in that smallest of instants between the time his fist started up and the time it landed, a beautiful woman's head with a cap of thick red hair cut like a young boy's popped out from the side of the stockinged man who was still screaming the last word: "NOW! NOW! NOW!" The woman was Jake.

Chapter Six

When he opened his eyes Eugene was still where he had dropped. For a moment, he remembered nothing, and did not know where he was. Then he remembered Jake and his heart quickened. He would not have been any more surprised if Pete's face had appeared looking up at him from under the man's shoulder. He felt for the envelope, found it, and tested its thickness between his thumb and forefinger.

"Don't worry, it's still there. We run an honest business here."

He sat up and turned. Jake was standing in the doorway in a long kimono not unlike the one the houseboy was wearing, except hers was black and the dragons breathed fire as red as blood. Eugene got off the floor and sat on the edge of the bed.

"How do you feel?"

"I feel all right."

She was slender, almost delicate, her collarbones insistent under pale, nearly translucent skin there where the kimono was open to a point between her breasts, which were small, the breasts of a young girl. Except for the nipples. The nipples were large under the silk robe.

She walked over to where he sat. "Here, this will make you feel better." She handed him another envelope. He opened it. Three one-hundred-dollar bills. "That's lagniappe," she said.

"Why lagniappe?"

"Your jacket's got heat under it. My trick'll always give lagniappe for leather and a pistol. He saw it, gave you lagniappe, then I had to get my knee pads and give him lagniappe."

"Sorry," he said.

"Don't be. I don't mind. My knee pads and lagniappe for a man come at an exorbitant price."

Eugene had heard the word *lagniappe* from the day he arrived in New Orleans. He had finally seen it spelled on the menu at Jacques' and asked how you said it and what it meant. It meant "a little something extra," and it was pronounced "lan-yap." He had written his daddy about it and his daddy had written back that he did not care how they said it, he did not have any use for a word like that. It did not seem to him that a natural man would want to say such a thing and he hoped that Eugene would not use it in front of womenfolk and that he would be careful around anybody who did.

Eugene had been breathing deeply and snapping his head from side to side in the old gesture from the corner of the ring just before a fight. He stood up and found himself steady. Not even groggy. He looked over at the open door of the bedroom.

"The guy still here?"

"Oh, no," she said. "He's gone. This is my place."

"A little wild with his money, ain't he?"

"He's a Louisiana politician," she said. "Enjoy what he gave you, it came right out of the public till. And don't feel bad about taking it. Even in a state like Louisiana, the public till is bottomless."

"What kind of office does . . ."

"You don't want to know."

"Right. I'm just not thinking very good yet. I saw the stocking, thought it was just kink."

"He's got kink enough, but the stocking was strictly business. He didn't get to be where he is by being a fool. He doesn't know you so he doesn't trust you."

"I've known a lot of people for a long time. I've never burned down anybody's proposition. The book on my tongue is that it's solid as rock and clean as snow."

"I knew that. But he didn't." She walked to the door and looked back. "Let's go into the living room."

She took a huge, uncomfortable-looking wicker chair, the back of which extended two feet above her head. He sat on a small sofa covered in some sort of hide he did not recognize, solid black, the soft hair still on it, and surprisingly comfortable.

"Can I get you something?"

"If it's no trouble, I'll have about three fingers of vodka straight up."

From a delicate-looking table beside her chair, she took up a gold-colored triangle with a cylinder big as half a pencil hanging from it on a chain. She struck the triangle and immediately the young Chinese boy glided into the room. "I believe you met Dong, my houseboy?" she said.

"At the door," Eugene said.

Dong put his hands together as if in prayer and bowed his head briefly.

"Bring the gentleman three ounces of vodka straight up and bring me a Campari."

When he was gone, Eugene said, "Dong?"

She smiled. "Orientals have names that sound strange to us. I can assure you that ours sound just as strange to them."

"I don't believe I'd want a houseboy named Dong."

"Dong's a fine houseboy and has the added attraction of being gay. A boy who was straight would make obvious complications, as would a woman, no matter how she swung."

"I shouldn't have said anything. It ain't none of my business."

Dong was back with the drinks on a silver tray. After he set down the glasses, he swayed silently out of the room. Jake took a sip of the Campari and said, "Let me give you some advice that is none of my business. You should avoid the word *ain't*. Sometimes you do, sometimes you don't. Nothing wrong with the word. A perfectly good word that nobody misunderstands. But it causes a whole string of negative judgments to check in with many people when they hear it. Same thing with double negatives. They are perfectly acceptable in many languages. But not in English. *No tengo nada* is fine in Spanish, but one mustn't say 'I don't have nothing' in English. I've got somebody I want you to meet and it may help you considerably if you'd let me help you clean up your grammar. It would not be difficult. You'd pick it up in a heartbeat."

"Seems like somebody is always trying to teach me something, and it always gives me a headache." Eugene finished the glass of vodka. "I know more people than I want to know already, and I ain't trying to impress nobody."

She laughed. "As you will, Eugene, as you will."

"You've lost your Brooklyn accent," he said.

"That was goof and spoof, this is business."

"What's your name, anyway?" he said. "If you don't mind me asking."

"You know my name. Jake. My last name is usually Smith or Jones. But to you it's Purcell."

"You had a mama and daddy who named you Jake?"

"My mother and father are best left out of it. I named myself, just as I certified myself. Would you like another vodka?"

The vodka had quickened his blood and lifted his spirits. He felt wonderful. "If you're offering, I'm taking. One more would get me just about where I need to be."

After Dong brought two more drinks, Eugene said: "Why do you give me your last name straight?"

"Because you're one of us."

He did not know exactly what she meant by that. He thought about it for a minute and decided he did not want to know. He sat sipping from his glass, and then something occurred to him and he felt his spine stiffen. He looked down at the slight bulge under his jacket.

"How did the guy with the stocking know I had a pistol on me?"

"He unzipped you. No, no, don't get hot. Just the jacket, only the jacket. I was there. He asked me if it would be all right. He wanted to unzip the jacket, to see you. That was all. For what reason, I don't know. I didn't ask, he didn't say. I wouldn't have let it go any further than that, not with you flat on your back and out of it."

"That was not part of the deal, the son of a bitch."

"You'll get over it."

"He won't get over it if I run up on him again. He'll need more than a stocking to cover his ass."

"All of this has been a little hard for you to understand, hasn't it?"

"I don't need to understand it. I haven't tried."

"I sensed that about you from the beginning, how nonjudgmental you are, or at least seem to be."

"I know somebody else who talks that way. It makes my head hurt, saying things like nonjudgmental does."

"Fair enough. But still you haven't said a word about me fucking that guy. After where we last met I would have thought you would."

110

"Jake, as the brothers say, I'm just trying to make it on over."

"Dong!" she called, a little network of wrinkles pinching in her forehead as though something had suddenly angered her. When the houseboy appeared, she handed him her Campari and said: "Take this away and bring me a vodka. And bring the gentleman another. As a matter of fact, get a bucket and bring the bottle."

Dong took her glass and hurried from the room in a rustle of silk.

"I've got a motorcycle downstairs. I'd rather not get on it if I'm on the outside of a half a quart of vodka."

"Then don't get on it. It's almost two o'clock in the morning. If you get drunk you can sleep here. There's a bedroom with a private bath. Nobody'll bother you, certainly I won't. And before you ask, which you probably would not, or wonder about, which you might, it doesn't mean anything, saying you may stay if you like. It's just that . . . that I rarely talk to a straight guy because they're all such assholes. But you're straight and you're one of us. That's unusual. I like unusual."

Dong brought the vodka bottle iced down in a champagne bucket. She walked over to the window and drew the drapes a little way back. "You can see the whole Quarter from up here. Two o'clock in the morning and it's just starting to cook down there. God, I love it. I wouldn't want to live anywhere else. Your accent tells me you're not from Louisiana."

Eugene didn't say anything. He was looking at the iced bottle and at his dry glass and thinking about the motorcycle and the ride back to an apartment that would be empty when he got there. He knew Charity would not be back tonight and Pete was already off and gone God knows where, probably with Tulip. He reached for the bottle as she asked again where he was from.

"I'd tell you but it'd bore the hell out of me. And you, too."

"Then don't. Do anything to me, but don't bore me."

She came away from the window and poured a drink into a

glass that Dong had left beside the bucket. She took off the glass in one long drink and poured another. Taking a quick sip, she said: "Fucking men is a way to support my life the way I want to live it. And it's a way to support my habit."

"Jake, I don't need to know this."

Eugene was already beginning to disengage himself from what she was saying. He could listen to it all night without letting himself think about it too much, but he didn't want to. God, he did not want to know this private shit about her. And habit? He had seen enough needle freaks to last a lifetime.

"Small talk in the small hours, man. That's all." She took another pull at the glass. "You wouldn't believe the shit I've done going back as far as high school."

"None of us would believe what anybody has done if we really knew. All of us do what we do because of what we see and what we believe we know," he said bitterly, thinking against his will about Budd and that last terrible night with him in the dressing room here in New Orleans. "And that ain't worth nothing. It ain't just shit, either. It can kill you, or else leave you so high and so dry you can't breathe." His glass was empty again and he realized that he was getting drunk, but he had had just enough to drink that he did not care.

In a quiet, bemused voice, almost as if talking to herself, she said: "In high school, I used to go out with a boy and on the first date I'd fuck him, fuck his eyes out, pretend I could not help myself because I had been in love with him for six months, but had not had the nerve to tell him. Talk all kind of shit to him, a lot of love shit and us forever shit, the whole time fucking him until his eyes cracked, all on the first date. And then the next day I wouldn't speak to him, wouldn't explain anything. Never go out with him again or even go near him again. I wouldn't even talk to him on the telephone."

It may have been small talk in the small hours, but to Eugene, it sounded like a confession. Or maybe she just wanted to talk and this was what had come into her head. Too, she was hitting the vodka pretty good, pulling the bottle from the bucket again, still talking in that quiet, bemused voice.

". . . and occasionally, I'd even go down on a boy, always on the first date, same shit as before, but I wouldn't fuck him, just go down on him but not get him off, always stop short of getting him off, and all the time promising through a hundred declarations of love that I would give it all to him the next night or that weekend or whatever. Then I'd never speak to the poor bastard again." She emptied her glass. Eugene thought she ought to be getting drunk by now, but her voice was steady as ever. "When I was in the tenth grade a boy killed himself. When I was a senior—it was just before graduation—another boy, the fastest boy on the track team and class president, tried it. He hung himself but the rope broke, and he didn't die. He only paralyzed himself from the neck down. Permanently."

She rose from the high-backed chair, and with a steady hand poured him another drink, and then herself one. She showed no inclination to stagger. Eugene thought that he might be her equal in a lot of ways, but she would beat him every time over a bottle. The damn girl could drink up some vodka. He had already decided that he was not going to get on the BMW and he was not going to call a cab, which he could easily do. He was going to stay right where he was and slide. He was to what he called the fuck-it stage of the evening. He came to the fuck-it stage often and he really didn't need alcohol to arrive at it. Other things could do it, but however he got there, when it came, he simply no longer cared about anything, or anybody, including himself, maybe particularly himself.

He still felt alert, his mind clear, but he knew he was drunk. And he could count on his fingers the times he had been this drunk in his life. What was going on with him these days? What before had been, if not easy, at least tolerable was becoming

. . . what? The only word he could think of was *scary*. Screw it, he told himself, your brain—as Pete would say—is wet. If it's wet, drown the motherfucker. He reached for the bottle.

"I did that stuff with boys all the way through college."

"Wonder you didn't get killed," he said. Nothing she could say now could touch him, not in the fuck-it stage of the evening.

"That occurred to me," she said. "I thought of that after the boy killed himself. If one of them would kill himself, why wouldn't one of them kill me? But it didn't stop me. Being a child, just being young, is a cruel, vicious time."

"Well," he said, his lips numb now, "it's one good thing about it. We grown up and veterans now, you and me, and we entirely happy and contented, totally on top of things."

"You got that right, Eugene, we've got our shit stacked up neat and nailed down tight, nothing if not happy and contented."

"And drunk. Don't forget drunk."

"For whatever it's worth, and it's not worth much, I suppose, I stopped that fuck-and-abandon routine with guys a long time ago."

"Definitely a step in the right direction."

"Don't make fun of me."

"Hell, I ain't making fun. Small talk in the small hours, remember?"

"I whore, but that's all," she said. "I whore to support my habit."

"Yeah, you said. But, you know. Whores. Bout the honestest people on earth. Seems like to me. You say to a whore, I want this and this, and she says this will cost you this, and this will cost you this. And that's it. The bargain's struck. I don't know. Just seems like to me. Caint say the same for pimps, though. Pimps will eat your lunch."

"I don't have a pimp. Nobody runs me on in the street or anywhere else."

"Good for you." He was reaching for the bottle, but kept missing it.

She got out of the chair and poured his drink. "Did you wonder how you happened to be here tonight?"

"Not really. I try not to wonder. It seems to give me headaches."

"I got Joey Q to call you. The sucker that was here tonight is over kink like a glutton over food. It doesn't matter much what the kink is, he'll go for it, at least once. So I thought I might as well throw the work your way. Besides, I had to talk to you."

"I don't know no Joey Q."

"You don't know him, but you met him. You've just forgotten."

"So you had ole Joey Q call me up. Wonderful. And I got a place for the money, I do have that."

"We've all got a place for the money. But don't you want to know why I wanted to talk to you?"

"You already said."

"Maybe I ought to wait until in the morning. After you've slept."

"You want to talk to me? Talk to me."

"Mr. Blasingame wants to see you. He asked me to talk to you."

"Well, shit. I'm a little wet and warped here, but I know I don't know nobody named Blasingame. Nope, no Blasingame."

"Yes, you do. I'll give you his card. His office is in the CBD."

"The ole CBD."

"The Central Business District."

"Oh, yeah, right. Never had much reason to go down there."

"You do now. Mr. Blasingame wants to see you."

Eugene raised his hands, palm up, and shrugged.

"Oyster Boy," she said.

"Ole Oyster Boy," said Eugene.

"He's Mr. Blasingame in the CBD."

"How's ole Oyster Boy's skin doing?"

"You ought to leave that alone. Mr. Blasingame's skin is no concern of mine. Or of yours."

"He still on a leash?"

"We're talking about a different thing here. He is considerably more than a man with bad skin on a leash. He is not what he seems. Or said another way, he's more than he seems."

"I'll drink to that. Nothing and nobody's what they seem. Confusing, ain't it?" He reached for the bottle.

"You've had enough of that. Come on, I'll show you your room."

"I think I'll just sit here and drink. Sleep on the couch."

"I'm afraid that won't do. Femally will be in sometime tonight. Should have been in long ago, the gorgeous, worthless bitch."

"Fe-male-ly."

"Rhymes with tamale, as in Mexican food."

"Good name, a great name, Femally."

"Come on, it's late. I'll bring the bucket. You can drink yourself to the moon if you want to once you find the bed."

Eugene stood up and it felt as if all the blood had drained from his head. "I do believe you gone have to give me a shoulder, Jake."

"I'll call Dong."

Eugene sat back down. "No Dong. Couldn't you give me just a little bit of shoulder?"

"If that's what you want."

He stood up and put a forearm on her shoulder and she started back across the living room, but after the second step he had taken, he remembered again what he had not remembered or even thought of in a long time: Budd's thick shoulder leading him back to the dressing room after those last terrible fights, with the crowd still cheering for the fighter who had defeated him.

He took his arm away. "I can make it by myself. I'm all right."

Chapter Seven

He awoke and for a long minute he did not know where he was. A strong sun beat against the heavy drapes at the window. He raised his head and saw his clothes hanging neatly from a silent valet at the foot of the bed. The holstered .38 hung on a peg below his clothes. He moved his hand over the sheet under the light blanket covering him. Satin. He lay very still. Gradually at first, and then all in a rush, the whole evening came back to him. Oyster Boy? Blasingame? Wanted to see him, Eugene, in the CBD? He moved his head on the satin pillow, and an intense pain burst behind his eyes.

He sat up on the side of the bed and looked down at himself. He was wearing bright pink silk pajamas with line drawings of trees and birds embroidered into them. The last thing he remembered was taking Jake's shoulder and thinking of Budd.

A strange, yet familiar, fragrance floated about him. He

raised his arm, turned his head, and took a deep pull of air through his nose. The sweet smell of body oil floated out of his armpit. Christ, sweet Jesus, where had he slipped to this time?

He got off the bed, stripped out of his pajamas, and went over the ocher-colored carpet to the bathroom, stepping carefully to avoid the stabbing pain behind his eyes. There on the marble shelf above the sink, arranged in a line, were a razor, shaving cream, after-shave, a packaged toothbrush, toothpaste, a comb and hairbrush, and finally a large bottle of aspirin. Thick white towels were folded on top of a small metal stool beside the opaque glass of the shower.

First things first. He got into the shower, adjusted the temperature of the water until it was as hot as he could stand it, and then scrubbed himself a long, slow time until his skin was free of the odor of body oil.

Out of the shower, he took four aspirin, rubbed the steam from the glass, and shaved. By the time he walked out to the place where his clothes were hanging, the headache was beginning to fade, and he was feeling pretty good. He was still shaken and very pissed about the pajamas and the bath oil. But what was done was done. The answers would come or they would not. He would break somebody's skull or he would not.

When he came out of the bedroom all the drapes of the living room had been opened and sunlight flooded the room, bringing the dark colors alive and making it beautiful in a way it had not been last night. Through French doors where one of the drapes had been pulled, he saw Jake sitting on a little balcony at a table with a cup of coffee, reading the newspaper. There was a high wrought-iron railing behind her.

Without looking up, she called, "Come on out. It's a beautiful morning."

He went onto the terrace and took a chair at the table. He had no idea what time it was, but looking out over the French Quarter, he knew it must be early because it was quiet as a graveyard down there. The drunks had all stumbled back to their hotel rooms and the bars were closed and shuttered.

Eugene said: "About last night, who . . ."

Dong came onto the terrace balancing a tray with a pot of coffee and a glass of orange juice on it. It was only then that he noticed that Jake was wearing the kimono that Dong had had on the night before and Dong was wearing hers. Eugene refused to let himself wonder why. But the fire flaring from the dragons did not look as good, as natural somehow, on him as it had on her.

"Gentleman like breakfast?"

"No," said Eugene.

"No trouble. Dong like make breakfast for gentleman."

"That's quite enough, Dong," Jake said.

Dong bowed and, smiling a sweet, sad smile at Eugene, left them alone.

"Sleep well last night?"

"Yeah, but . . ."

"You're a cheap drunk," she said.

"I got a fucking question. I got up in a pair of silk pajamas and smelling like I'd slept in a pile of gardenias last night. Goddam body oil. How?"

"How what?"

"I don't sleep in pajamas and I never used body oil in my life."

"How'd you know it was body oil, then?"

"I know."

"And I know you were on your ass by the time you got to the room. I left you dressed as you are now. The pajamas were folded beside you on the bed. Not I or anyone else bothered you. What's the last thing you remember last night, just out of curiosity?"

He looked out across the French Quarter. "Starting to the room. And taking my arm off your shoulder."

She looked back at her paper. "Drink your coffee. I once

went to sleep in The World Under Hotel in the Quarter and woke up in Houston in the Holiday Inn. Lost everything between New Orleans and Houston. Compared to that, pajamas and a little bath oil's nothing." She folded the paper in half and looked over it. "By the way, Femally got home this morning. If she happens to come out, which she probably won't, don't mention her name. If she wants you to know it, she'll tell you. She's sensitive about it. I wouldn't want her to know I told you."

"Unusual name," said Eugene. "But I don't . . ."

"It's spelled F-e-m-a-l-e."

"But that's female."

"The story she gives me is that her mother was not too quick, saw it on the birth certificate, and thought that was her name, so that's what she's called today. My own feeling is that her mother was quicker than Femally ever knew. Because she is some kind of female."

He started to ask why she did not change it the way Jake had obviously done if she did not like it, but he had already asked questions that were none of his business. Waking up dressed in pajamas and smelling of gardenias had rattled him. Things had changed up on him. His head was beginning to ache again. He drank the coffee and poured another cup. He hoped his motorcycle was all right. Leaving it parked in the street on the edge of the Quarter all night was not a good idea. It could easily be in somebody's van by now.

Jake folded and put down her newspaper. "How much do you remember of what we were talking about last night?"

"All of it."

"Good." She brought out a card from somewhere in the folds of her kimono and pushed it across to him.

Across the card was printed J. ALFRED BLASINGAME ENTERPRISES, and under that an address in the Central Business District.

"I don't think so," said Eugene.

"You ought to go see him," she said.

"You forgot. I saw that movie. I even acted in it."

"You saw Oyster Boy. You've not seen Mr. Blasingame."

"I don't believe I want to talk about it. I've got to be going, anyway."

"That night we met on St. Charles, would you have thought that you would get drunk with me in my living room and then have coffee with me the next morning? Of course not. But it has not been an unpleasant experience, has it?"

Eugene thought again of the pajamas and the oil, but he decided to leave it alone. "It's been nice," he said.

"Then put . . ."

She stopped because there appeared in the doorway leading onto the terrace a black woman in a peach-colored dressing gown. Eugene thought she was the most extraordinary woman he had ever seen. Her hair was pulled to the side and held in a tight twist. Her eyes were enormous and her cheekbones high and flat. The dressing gown was short enough to show narrow ankles and diamond-shaped calves. She was the blackest black woman he had ever seen and by far the most beautiful woman of any race he had ever seen anywhere. Her face was smooth as burnished wood, not even a wrinkle about the corners of her eyes, though it was obvious to Eugene that she was a good deal older than Jake, and her teeth were impossibly white in a mouth that was full-lipped and smiling. Such a smile on a man would have given Eugene a minor jolt of anger because it was clearly and insultingly mocking. She looked at Eugene for a long moment before turning her eyes on Jake.

"Wake me up at four-thirty, love. I've got places to go and things to do." Her voice was heavy with the deep South, but lilting, too, and mellow. "Don't have the Chink do it. I want your sweet hand on me when I open my eyes. I'll be starved when I wake up. Do you think you, I mean *you*, could cook me up something outrageous?"

"I'd love to, just tell me what you'd like."

"I'm gone leave that to you. Just make it outrageous and make a lot of it."

"Four-thirty with my hand on you and outrageous on the table," Jake said.

"Solid," Femally said.

As she turned to go, Eugene did what he ordinarily would never have done. Maybe it was her smile, or maybe it was how her eyes had dismissed him. He said: "My name's Eugene Biggs. I'm a friend of Jake's."

Femally stopped, half turned in the doorway, and looked at him as if she were seeing him for the first time. "No shit. Now that's a fine name. You pretty, too."

"Thanks. How you doing this morning?"

She arched her penciled eyebrows. "Me? Just lingering, white boy. Lingering like a fart in a phone booth." And then she was gone, trailing soft, deep laughter.

When she was gone, Jake picked up her paper, but almost immediately put it down again. Her eyes were shining with what could only be tears. Eugene felt his heart go heavy in his chest. He wanted to comfort her, but did not know how. Her lower lip trembled, then steadied.

"In case you didn't figure it out," she said, looking out over the Quarter, "that was my habit."

He felt that if he said anything kind to her, she would break down completely and that was the last thing he wanted. "Habits," he said. "Like opinions and assholes, everbody's got one."

She brought her thin, long-fingered hand down on the table sharply. "Goddam the world!" But then she grew calmer and looked up to meet his eyes. "There ought to be a law against the sun rising and setting for you in somebody else."

"I guess," Eugene said. "She didn't think I was business, did she?"

"She knows I don't have coffee with the tricks I turn." She was silent a moment and then: "Did you mean it about being my friend?"

Eugene thought about it. "Without lying to you, I can tell you I ain't your enemy."

"Good enough," she said. "Then take the card."

"I'm not going to see this guy, Jake."

"Just take the card. Put it in your pocket. Take it with you."

"Why should it matter to you one way or the other?"

"It does. I owe Mr. Blasingame. I can tell him you took his card away with you. I can tell him that, at least."

"Lie to him."

"You don't lie to Mr. Blasingame."

Eugene picked up the card, turned it in his hand, and put it in his jacket pocket. It was not going to cost anything to carry it away with him.

"That's better," she said. "That's much better."

"Thanks for the bed and the business. I never run into that kind of money before."

"And you won't ever run into it again either, unless you run into a politician on the take. Very generous people."

"I've got to go, Jake. As we say in South Georgia, I've got hogs to call and chickens to feed."

"Is that where you're from, South Georgia?"

"That's where I'm from."

"I'm from Mississippi. We'll get together sometime and talk about chitlins and incest and other southern contributions to culture."

"You got a deal."

"Dong will show you out. I just want to sit here for a minute."

"I can find the door. Later on, Jake. Another time, another place."

"Later," she said, picking up her newspaper again.

Halfway across the living room, Dong was at his elbow. "Gentleman sure not like breakfast? Take only minute."

Eugene said: "Go somewhere and chill out, Dong, you drooling."

"Did you like the pajamas when you woke up wearing them this morning? Was the body oil satisfactory?" Dong had lost not only his smile but also his Chinese accent.

It did not surprise Eugene. Nor was he angry. If he felt anything, it was a deep melancholy that this was only part of the price of doing business.

"I ought to take you in the back room and beat you awhile, you slant-eyed son of a bitch. I wouldn't even close my hand. It'd be like beating a baby to death." But his heart was not really in it, and this seemed to give Dong courage.

"Jake would not like that at all. You had best think about where you are."

"Jake don't mean nothing to me."

"Femally's magnum would mean something to you, country boy. You fuck with me and she'll come out with that cannon or else upside your head with a straight razor."

Eugene unzipped his jacket to show the .38. "I ain't worried about her either."

Why didn't he drop this? This was all a game. It meant nothing to him. He did not know why it meant nothing, either. Where was his anger? Where was his outrage? The anger was there last night and it had been real enough when he had found out about Jake's trick unzipping his jacket. What had taken it out of him?

"You're slow this morning, country. Really slow. Why didn't you check your piece? I took the rounds. It's empty."

Eugene walked to the door leading out into the hallway and stopped. As he knew he would, Dong followed him, only now he was seriously baiting Eugene, his voice mocking. "There is really no need for this kind of rank aggression, all I did was undress you and bathe you and touch you a little."

Eugene reached behind him without even looking and wrapped his right hand in Dong's kimono. With his left hand he opened the door, and then snatched Dong out into the hallway. With the door closed, he jammed him against the wall, lifting as he did until only Dong's toes were touching the floor. Dong opened his mouth to either say something or scream, and Eugene took him by the throat with his left hand, gripping until Dong's eyes were starting from his head and he could no longer breathe. Dong jerked his head wildly, looking both ways down the empty hallway.

"Looks like everybody's still asleep, sweet thing. Say whatever prayers you little yellow motherfuckers say because you're gone. I'm going to leave you dead on the floor."

Dong beat his heels against the wall and flopped around, his arms and legs snapping in little spasms. The thought of actually killing him passed idly through Eugene's mind. He knew it would be no more to him than killing a turtle on a mud bank, or butchering a catfish while it was still alive. Dong's eyelids were dropping over his eyes, eyes in which the light had gone out. His face had turned from the color of gold to dull purple.

What shocked Eugene was not what he was doing but that Dong was dying against the wall and that he simply had no feeling about it one way or the other, not even the mildest anger. It shocked him enough that he released him and let him slide heavily to the floor. More out of curiosity than concern, Eugene bent to check if his breathing was going to start again. It did, shallowly at first, then in racking, choking gasps. Eugene walked to the elevator and never looked back.

At the curb, his motorcycle was just as he left it, except that he had a ticket. He slowly and carefully tore the ticket into small squares and dropped the pieces into the gutter. He got on the bike and rode back up Canal Street, nearly empty now of traffic. He was filled with a kind of wonder at what he had just done, and more, wonder at how he felt about doing it. And then as he rode through the quiet morning, he heard Jake's words again as plainly as he had heard them the first time: "You're one of us." And he knew as soon as the words came to him that they had never left him for an instant since he had heard them, that they had been somewhere in him alive and waiting to be heard again.

"What did that bitch think she was saying?" he said aloud into the wind rushing past his face.

And he was suddenly afraid, afraid as he had not been in a long time. He told himself that he was only tired, that it was the effects of the vodka that he had drunk last night. But still he could not shake what Jake had said to him. He cracked the throttle with his right hand and with his left brought out the card she had insisted on his taking. Anything to get his head away from where it was. He checked the address and roared through the deserted streets, laying the bike so low at corners that he was getting sparks off the chassis from the street.

In the Central Business District, he found the address. Thirteen floors up was a huge sign in copper-colored letters that said: J. ALFRED BLASINGAME ENTERPRISES.

He thought: I don't know what game ole J. Alfred plays, but he must be damn good at it.

He ripped the card in half and left it at the curb. There was no place to go but back to the apartment. He did not think Charity would be home, but even if she was, he planned to try to get some more sleep. He felt very tired. He stopped off for a big breakfast of ham, eggs, home fries, biscuit, and milk. He would rest better with something on his stomach.

Charity was not there when he got back. He lay down on the bed thinking he would fall right to sleep. But as soon as he closed his eyes, instead of hearing, he saw Jake's words in copper-colored letters, apparently written on nothing, and floating impossibly across the sky: YOU'RE ONE OF US.

He got out of bed and went to sit at the little desk Charity had bought for him. He thought about writing his folks a letter, but he knew he could not, because now the previous night and all performing nights before it—every bit of which he could remember in terrible detail—were focused around Jake's words. He didn't know what to do with himself. The immediate problem was getting through the day feeling the way he did.

As he turned to get up his gaze fell upon the two-drawer filing cabinet with the heavy padlock on it. Charity had him locked up there. It was not the first time the thought had come to him. He had watched her taking him off the Dictaphone, watched her putting into the cabinet the stacks of paper that spewed out of the printer. And further, he had watched her work at the word processor for long hours without the Dictaphone, and he had known, known for a long time, that she was writing about what he had told her.

He had never lied to her; that was not part of the deal. And he had never asked to see what she had in the filing cabinet, because that was not part of the deal, either. But he had thought of going through the lock and finding out what she had in there. And that was precisely what he was thinking of doing now.

Did she know something about himself he did not know? She was, after all, working on a PhD. He hadn't known anything about university degrees when he met her. If he knew anything, he simply knew that some people graduated from college and some did not. He did not know you could do it more than once. But you could, and she had already done it twice, and was on the edge of doing it again, of going as far as you could go in the university. There were no degrees left that she could study for. She was totally educated after this one, filled up to the brim.

She had to know something about himself he did not know. She was fucking educated. And going through the lock was no problem at all. It would be for him, but it would not be for the guy he had heard Pete talk about, one of Pete's friends, recently released from Angola prison.

Eugene dressed quickly and rode to Prytania Street where Pete had a little apartment over a shoe repair shop. When he knocked, the door was opened by a girl he almost did not recognize in a housecoat and wearing no makeup. The other times he had seen her, the only thing she was wearing was makeup. It was Tulip.

"My name is Eugene Biggs," he said. "I need to see Pete."

"He's asleep, but come on in. Pete talks about you all the time. You were a fighter, too," Tulip said.

"Not for very long, but for a while." Eugene came directly into a tiny room hardly bigger than a closet and sat in a ladder-back chair. The only other place to sit was a beanbag in a corner.

"I just made coffee. Can I get you a cup?"

"Be great," he said. He didn't want the coffee but it seemed impolite not to take it.

Tulip went into the kitchen. It opened directly off the room he was in and was so small that in the space between the refrigerator against one wall and a two-burner stove against the other, Tulip hardly had room to turn around. She poured him a cup of coffee and brought it in.

"I'm sorry, I didn't introduce myself," she said. "My name's Josephine Apolarus. Pete and I had breakfast in the wee tiny hours and I was just cleaning up before I . . . well, I was just cleaning up."

When she bent to the chair to hand him the coffee, she held her robe together at her breasts. Eugene remembered her breasts, full and high, and remembered her naked on the floor with the teddy bear there at Maxie's.

"It's good to meet you," he said, "and I'm sorry as . . . I'm sorry to come in here so early. I knew Pete would be asleep. Why don't you just let me go back and speak with him? I won't be a minute, and I'll be out of here."

"Sure. I know how far back you guys go together. He said when you were good, you were very, very good."

Eugene stood with his coffee. "Yeah, that's probably true. But I stopped being good early."

"I can run behind that," she said. "We all stop being good sooner or later. It ain't nothing but a thing."

There was a hard brittleness in her voice that had not been there before. They stood for a moment, before Eugene went back to see Pete. Eugene wondered if Josephine knew that he knew her as Tulip and had seen her work. For some reason he could not have named, he hoped she did not.

The room was airless where Pete slept naked, his black skin covered with a thin film of sweat, and he was snoring softly. The bed was two mattresses, one on top of the other, on the floor. There was only one window and it was covered with a blanket. Pete's outstretched arm was only inches from a half-empty whiskey bottle.

Eugene tapped Pete on the bottom of the foot with his boot. Pete opened one eye and groaned. "Awwww, man, shit!"

"I need a favor," Eugene said.

"What time is it?"

"Probably ten, a little after."

"Get away from me. I just got to sleep. I wouldn't do my mama a favor now."

"I ain't you mama, cocksucker. I need a favor and I need it now. You going to get your ass up or what?"

Pete sat up on the bed and groaned again. He rubbed his eyes with his left hand while his right hand groped for the whiskey. Eugene picked up the bottle, opened it, and handed it to Pete. He raised it and drank.

"Tulip still here?"

"She said her name was Josephine, and yeah, she's in the kitchen."

"I like Tulip better. And if she's still here, why ain't she asleep?"

"She's trying to clean up a little, man. Why don't you take that blanket off the window and open it? It smells funky back here."

"Noise, man. Light and noise. Enemy of the daytime sleeper. And it don't smell funky." He took another short pull at the bottle.

"All right. How about stinks?"

"That's what Tulip says."

"You ought to call her Josephine. That's what she wants."

"And what is it you want here in the middle of my night?"

"I need to get through a lock. You said you knew a brother just out of Angola who could walk through locks."

"Any except them at Angola. But I'm not interested in a rap for B and E."

"Don't talk shit. Ain't nothing to do with breaking and entering. This is a thing at my apartment. A padlock."

"You woke me up for that?"

"It belongs to Charity but she's not there. I'll pay and it's nothing but a steel filing cabinet. I'll pay big time."

"Man, what kinda weird shit you into now? No, don't tell me. Let me have another little taste of this and then you drop me over Jackson Avenue. You wait at your place. We not there thirty minutes after you drop me, we ain't coming. What do I tell my man you paying?"

"Tell'm I'm paying what it costs."

"He'll like that," Pete said, pulling on a shirt.

"You think he'll be home?"

Pete smiled. "He's just like me, K.O. He works nights."

"The name's Eugene."

"Sure it is."

After he was dressed, Pete came out into the kitchen and walked up behind Josephine at the sink and kissed her on the neck. "Leave that stuff alone and go get some sleep."

She turned, smiling, and put her face against his chest.

"I wanted to get the place cleaned up before I go. I can sleep anytime."

"Sleep now," he said.

"Are you going out?"

"I'll be back in a heartbeat. And I want to see you in bed."

"Promise?"

"Don't talk like that in front of this guy." He hooked a thumb toward Eugene. "He ain't nothing but trash, not nice like us."

"Wake me when you get back," she said.

Pete brushed her lips with his and was through the door ahead of Eugene. At the curb while Eugene was unlocking the helmets, Pete said: "What you smiling about?"

"You want the truth or you want a lie?"

"A lie'll do."

"I'll give you the truth anyway. That back there in the kitchen reminded me of my mama and daddy."

"Everything reminds you of you mama and daddy, man."

"I don't know what you doing, Pete, but whatever it is, keep on doing it. You might fuck around and make something good for yourself yet."

"Let's move it."

When he dropped Pete on Jackson, Eugene said: "I guess this guy's got wheels. You won't have to mess with no cab."

"With the tools he's holding? He's got a van."

Eugene rode back to the apartment and sat very still in a chair watching the filing cabinet until he heard a rap on the back door. "It's open," he called.

Pete came in with an albino black who was no bigger than a jockey. He was wearing a purple shirt, electric blue slacks, and yellow shoes. He had a pale red Afro, the color of diluted blood. The Afro had bits of lint in it. All the teeth he showed were gold, and a huge gold cross hung from a gold chain around his neck. His hands seemed too large for his body and were thin and long-fingered and constantly in motion. He carried a black leather bag that looked like it ought to belong to a doctor.

"Eugene, this is my friend Great."

"As in Alexander the. Can you dig it?"

132

They shook hands and Eugene said: "I like your flash, Great."

"You wear Angola gray for six years, forty-two days, ten hours, sixteen minutes and fifty-two seconds, you be buying street corner flash too."

"It's a wonder somebody don't rob his mouth," Pete said.

Great said: "These golden snags be from a better world and a better time."

"Pete," Eugene said, "get on the back door for Charity."

"Make it fast," said Pete. "I got business."

Eugene walked over to the filing cabinet, laid a hand on it, and said, "This is it."

Great set his bag on the floor. He lifted the padlock and let it lie in his wide, pink palm. "What's in it?"

"Papers."

"Negotiable?"

"I don't know nothing about negotiable. It's just pages of typing that's no good to anybody but me. What's the freight going to be?"

"You got a free ride. I was gonna take a bite if it was cheese or stones or something, but paper? It's free, Knockout." He was working now with a tiny metal pick, thin as a razor blade. "In the first place you don't think Great would come over here to do a padlock for anybody else but you, do you? Sending Great to do a padlock is like sending Muhammad Ali to fight a cripple. Only for you, Knockout."

"Pete runs his mouth too much."

"Pete only said it was a favor to The Knockout Artist. He had to or I wouldn't of come. But I already knew you, or leastwise the book on you."

"You thinking about somebody else," Eugene said.

The lock popped open in Great's hand. He turned and looked at Eugene. "You a legend in Angola, man. A freaking legend. What a proposition you got working. Beautiful."

"I don't . . ."

"Mind if I look through these drawers?" Great said, rifling through the printed pages in manila envelopes. "Got to make sure I ain't opened Fort Knox here and walked away with nothing but you admiration. But you right, Knocker, ain't shit here."

"Would you be willing to sell me that pick?"

"Wouldn't do you any good if I would, which I won't. You couldn't do a thing with this pick. It takes me."

"I may have to get in this thing again."

Great looked into his bag, then rattled around in it with his hand. "Oh, yeah, Lord. I just make you a key." He drew out a file and a blank key.

"How long will this take?"

"Four, five minutes max." Great was already filing on the key, held by a pair of vise-grip pliers. He glanced at Eugene briefly. "Down Angola, everybody got a proposition, or wish they did. I seen propositions in the joint you would not believe. Had a guy in there wore his head wrapped up and called hisself Rasnamd Narashi Ji, but he was just a nigger from Jackson, Mississippi. His proposition was he could put a ball bearing or a peanut or a marble in his asshole and stand there real still and draw whatever he put in there all the way up through his body till he opened his mouth and there it was on his tongue. We stood the dude in the shower, looked in him good, ever hole he had. Was clean. We hand him a steel ball bearing from the machine shop. He pops it in his ass and after a while he opens his mouth and there it is. Same ball bearing. We'd marked it. Look in him again. He's clean. Made him big money with that proposition out here in the free world. Knocks me out what you can beat people with. Sometimes wonder why I got to thief to make it on over." He tried the key. "Not yet. Couple more licks with this file, we home free. I could tell you a hundred like Rasnamd Narash Ji. All bullshit naturally but looking real as a genuine silver dollar." He tried the key again and the lock popped open. He handed it to Eugene. "You in business now."

"Mine's not bullshit," Eugene said. "I do what I say I do."

"Don't we all," Great said, showing his solid-gold smile.

He closed his bag and the two men shook hands.

"I'm much obliged," Eugene said. "Here, take this." He handed Great a bill. It was a hundred.

"Be glad to take it. Be glad," said Great. "But I wouldn't of asked for it. Be in touch."

"Right."

In the kitchen Eugene thanked Pete and Pete told him to try to keep his troubles between working hours and not to roust him in the middle of his night.

Eugene went back into the room and stood very still in front of the filing cabinet. For a long time a lot of people thought he was running a scam—a few still did—pretending to knock himself out. But, he had always told himself, he would have been incapable of such a thing. Dollar value for dollar paid. He could rest easy about all that. Every time, what he had said he would do with his fists was what he had done. It had been the truth, in the same way that everything he had told Charity was the truth. It would have been easy enough to tell her lies. At the very beginning, it had occurred to him that he *could* do it, but it had never occurred to him that he *would* do it. That was not the deal he had made. He looked at the open drawer of the file. What he was about to do was not part of the deal, either. But things had changed up on him. He sighed. He had done worse things before and no doubt would again. He didn't have to like it; he only had to do it.

He reached in and pulled out the first manila folder. On the tab across the top of the folder was printed: CASE HISTORY OF EUGENE TALMADGE BIGGS, A.K.A. KNOCKOUT, KNOCKER, K.O.

The bitch. The lying bitch. She had promised him that his name would be nowhere in the file, only his life, or as she put it, his essence. The rate of his heart jacked up a notch and he felt as he used to feel when he was just about to leave the corner for a fight, when he was about to go out and test himself.

But in those days Budd was behind him. Here, he was all alone. He opened the folder and read the sentence at the top of the page.

Men are so necessarily mad that not to be mad would amount to another form of madness.

PASCAL

He read it again. It made no sense. He was not mad at anybody. Then he remembered the dictionary and remembered also that Charity had reminded him that that was where one always went in a confusion about words. He took the folder with him to her desk, where he opened her dictionary and looked up the word *mad*.

1. mentally ill. insane . . .

He slammed the book shut and dropped it on the floor. He read the sentence again, but reading it this time: Men are so crazy that not to be crazy would be another kind of craziness.

Chapter Eight

Eugene sat on the floor with the printed sheets in his lap and a quart of vodka beside him. He had stopped worrying about or even thinking about Charity coming to the apartment. He had gone out for the vodka after reading, below the quote from Pascal:

> I drink not from mere joy in wine nor to scoff at faith—no, only to forget myself for a moment.
>
> OMAR KHAYYÁM

Will Biggs become an alcoholic? Probably. Perhaps, by some definitions, he already is one. He must take alcohol after every "performance" or "gig" as he calls it when he knocks himself out. Never heard him say "knock myself out." He hides it under all manner of excuses and disguises. It's a way to have the life he wants, he says. His family has to have the money, he says. But his problem is the same as the problem everyone has the world over, the problem of death. What to do about it, what to think about it, how to explain it to ourselves in such a way as to keep on living until it comes, but in the meantime (until it comes), trying to deny the reality of death. Biggs has solved it in his own unique way, or rather the solution was forced upon him. Every time he knocks himself out, he dies a little death, descending much lower than the son of Hypnos, Morpheus, the god of dreams. But he is resurrected. He resurrects himself. And his audience participates with him every step of the way. His audience is fulfilled. The audience itself, in that instant of death denied, escapes the ultimate reality of death. He (Biggs) is the quintessential hero. Here in the twentieth century is a living example of Man's recurrent, archetypal pattern for working out the Role of the Hero and at the same time denying the inevitability and finality of death. I must read all the literature from primitive and ancient times of the hero descending into the world of the dead, going back and walking through the spirit world, and then returning alive. Note: Check the mystery cults of the Eastern Mediterranean which I seem to remember as cults of death and resurrection. There is no evidence to suggest that Christianity evolved from these cults but plenty of evidence to suggest that Christianity was in competition with them. Biggs' audience has the same emotional bath, a bath that lays claims to immortality, that Christians have on Easter morning. "Hallelujah, He is risen! Even now He has healed Himself of death! All praise!" Has there been any religion reaching back to the beginning of recorded history that has not had the same primary goal and objective: how to bear the end of life? No! A thousand times, No! Any wonder that Henry James called death "the worm at the core" of man's existence? None.

And then anality. As with many emotionally crippled people, Biggs cannot forgive himself for having an anus. Literally, his asshole is killing him, or may be killing him, depending upon how much longer he continues to knock himself out. Biggs has never come to grips with, been able to cope with the burden man has always had to bear: He has a mind that is capable of the most sublime thought and at the same time a stinking hole that drops shit. Who said, "The highest king in the land sits on his ass"? Can't remember. How to justify Beethoven composing the

Ninth Symphony and farting at the same time? Difficult at best. I will make it clear that Biggs had a sublime dream. Without education, but not without a certain acute, even poetic sensibility (or so it seems to me), he managed to invest the brutal, indefensible, so-called sport of boxing with a sense of the sublime. And not only was that dream taken from him but taken in the most brutal fashion. Don't mean starting to lose, having his chin taken away from him, as he says, at Madison Square Garden. All fighters (all people?) start to lose sooner or later. No, it wasn't losing. It was Budd, his manager.

Eugene looked away from the page, back briefly, and then away again. The appearance of Budd's name on the page startled him with something as sharp and quick as electricity. He took a long drink from the vodka bottle, but the liquor was not helping very much. All that stuff about his asshole and shit and Beethoven farting and walking among the dead had kept him cold sober behind enough vodka to have made him drunk. He looked back and read the last sentence again.

It was Budd, his manager. Biggs loved him. Budd was in fact as much his father as if he had sired him. Biggs had to leave home, had to leave his father, whom he genuinely loves and admires, but having left home and his blood father, he found another in Budd. And when Budd, abruptly and in utter cold blood, rejected him, abandoned him here in New Orleans . . .

Eugene threw the pages away from him as if they had been snakes. A fear and horror such as he had never known seemed to charge his very blood. He reached for the vodka, but his trembling hand in an involuntary spasm knocked it halfway across the room. How did she know what she could not know? He had never told her anything about Budd abandoning him.

Or had he? Surely he would have remembered. But there were times, many times, when she said or did little things that made him think that she had wormed her way into his most secret heart, got to a place too sensitive and private for even him, himself, to bear. For a long time he watched the pages where they scattered when he had thrown them. He might as well go on with what he had started, shouldn't he? What difference did it make? He was about to get up to retrieve the papers when he saw the label on the thickest folder of all those that were stacked beside him. The label read: CONFESSIONS OF EUGENE TALMADGE BIGGS, A.K.A. KNOCKOUT, K.O., KNOCKER. Confessions? Confessions, for God's sake. What had he to confess? He lifted the folder, thicker and heavier than any book he had ever held, including the Bible. He opened it at random and knew immediately what it was: her questions and his answers. All done in the bed. All done locked together and sweating. His eye fell to the middle of the page where he had opened the folder.

C: But you might?
E: I . . . I . . . never . . . don't think about it.
C: Yes you have. Say it. Say it for Charity.
E: Yes. God. Oh, God, I . . . Yes.
C: Have you ever been touched that way by a man?

He slowly closed the folder. The fear and horror he had felt earlier slowly drained from him. He felt as he had felt when he looked into the mirror in the dressing room after he had knocked himself out the first time, when Budd had just left him forever. If this was it, this was it. He felt burned clean. This folder, these pages he was holding, was himself. He held himself in his hands. All of himself, the Eugene Talmadge Biggs that he had never intended to tell to anyone, not even himself. Maybe particularly himself. He slowly opened the folder again to the first page, cold with a numb acceptance.

He read. And he read. All through the long afternoon he read page after page of transcript. Most of what he read, he remembered saying. But much of it he did not. And stranger

140

still, most of what he did not remember, he would not and could not have brought himself to tell to anybody. And if some of what he had said had come into his head as thoughts, he would have dismissed them. For reasons he did not understand, he could not dismiss them printed on the page. He recognized them as true. He recognized them as himself.

When it grew too dark in the room to see, he moved to the little desk and turned on the light and went on reading. It was all there. His childhood. How it was with his mother and father and his brother, Edsel. His early country boy's sexual fumblings with animals. Leaving home, having to leave home. The boxing. The thirteen whores. And more. His secret longings and hopes and failures and shames. And many thoughts that frightened him.

It was all there, but not in sequence. Rather, it was jumbled and halting, bits and pieces, but long raving monologues too, monologues of which he would have thought himself incapable. He would have in fact denied even the possibility of such expression (confession?) as his own if they had come any way but the way in which they were coming, off the printed page, direct from the bed into the Dictaphone, and then through her fingers, each word, onto the printed pages he slowly turned there under the lamp on the desk.

And then he came upon a curious passage. He turned a page and his eye skipped down to it immediately because it was not written the way everything else had been written, in questions and answers with an E and C to begin each thing he said and each thing she said or asked. There was no C or E. And he read:

God of us all I understand his pain and torment. Rejection is a terrible thing to bear. Yes. I know. Oh do I know. No wonder he knocks himself out. Much of it has come from the hatred he feels for himself. Self-loathing. Yes. Why should I not understand? When he was betrayed by the very one he had counted on to direct him, told by that very one that he could no longer do what he most wanted to do, that he was not capable enough, worthy enough to continue, it was like death. It was death, or at least as much of death as one can know and still continue to live. But his way—punishing himself, hating himself—will not be my way. Henry Goldman can give up on me, but I will never give up on myself. I . . .

Eugene looked up from the page. All of this was obviously Charity talking to herself, and he knew the name Henry Goldman. Dr. Henry Goldman. Eugene had heard endless talk about him. Praise for his brilliance, how much he thought of Charity's future in the field of psychology, how much he had helped her, how much he had taught her. He was her major professor, the director of her dissertation, the book she was writing. The phrases "major professor" and "director of her dissertation" meant little to Eugene, and he would not even have remembered them if he had not heard them repeated so often. Well, what did all this mean if . . . He looked back at the page, his eye skipping down a few lines, picking it up in mid-sentence.

. . . and say I'm brilliant, a researcher without equal in the graduate school, and say at the same time that I need therapy, professional help, as they like to put it. Professional? I *am* a professional, god-dammit! Therapy? Bullshit. Rank nonsense. And Goldman! Did he come to my defense? Help me? He showed me nothing but his back. Abandoned me in the same way Budd abandoned Biggs. But I will not punish myself. I will not hate myself. No, Biggs' way will not be my way. They can have their university and their degree. I will do the work that is mine to do. And someday—someday— they will try to honor their university by giving me an honorary doctorate. And I will spit in their faces. Courage. Courage will take me where I need to go. And, thank God, I've been able to keep it all from Daddy.

C: And how many times? Tell Charity how many times.

Eugene closed the folder and stared at it for a long moment. He was confused but not shocked. He felt as though nothing would ever shock him again. At the center of himself he was cold and calm. Charity was not in school. And all that he had seen of her, all that he had known of her, all that she had said to him had been a lie. He had been revealed to himself, revealed to himself in an unthinkable and unbearable way, but she was protected. There was something between her and the rest of the world that had kept her safe. It was . . . What was it? He knew he knew, but he could not say what it was. And then he had the word. *Crazy.* And with the word came a piece of conversation, her voice, his voice.

"I worry sometimes that you might be just a shade crazy."

"Crazy? A meaningless term. I don't know what crazy is. Do you know what crazy is?"

"You damn right I do."

He got up from the desk and took the folder across the room to the filing cabinet. He reassembled the sheaf of papers he had thrown across the floor, being very careful to get them all in perfect order. He put the folders, one beside the other, just as he had taken them from the cabinet and put them all back in the drawer. He slipped the padlock back into the hasp and snapped it shut. He took the key Great had made into the kitchen where he wrapped it in a paper towel and dropped it in the trash. He would never need it again. He wished to God that he had never got it in the first place, but what was done was done. He could only do the next thing in front of him, but he had no notion at all what that thing was. He went back and turned out the light in the lamp, and lay on the bed and stared up at the dark ceiling. He did not know how long it was—it could have been hours or only minutes—before the phone rang.

He found the phone in the dark and put it to his ear.

"Hello. Hello? Eugene?" It was Charity.

"Yeah, it's me."

"You all right?"

"Any reason I shouldn't be?"

"You just don't sound . . . you sound funny."

"Then laugh."

"Are you angry?"

"Is there any reason I should be angry?"

"You're angry with Charity, aren't you?"

He realized for the first time that her habit of calling herself by her first name made her sound like a child, like a spoiled child. But now knowing what he knew, it made her sound like a spoiled, crazy child.

"No, I'm not angry with you." And that was the truth.

"How did it go?"

"How did what go?"

"The gig."

"You mean when I knocked myself out last night?"

There was a silence on the line. And then: "Did everything go all right? You don't sound like yourself. You sound . . . different."

"It's the middle of the night, Charity, I'm in bed."

"You don't sound as though you were asleep."

"I didn't say I was asleep. I said I was in bed."

"What *is* the matter, Knock . . . Eugene?"

"You want something, or is this just general conversation?"

"I'm coming home. Do you want Charity to come home to you?"

"Do whatever you want to do. Come if you want to."

"Will you be all right without me?"

"I'll be just fine without you."

"Well, I am right in the middle of some research and I'd rather not quit just now. So if you're sure you are all right, I'll see you tomorrow."

"Whenever."

"You try to get some sleep. You sound drained, exhausted. Sleep sweet, now."

"Yeah. And good luck with your research."

"Why, thank you." Her voice shifted in pitch. "You've never said that before, never once."

"No, I never have."

"Well, I appreciate it, I do. And good night."

He hung up the phone and eased his head back onto the pillow. When he woke up the next morning, he realized that he had slept soundly, dreamlessly, which surprised him. He even felt good, almost elated, which surprised him even more. The sun was bright against the window. He glanced at the clock. Quarter of nine.

He got up, stripped out of his clothes, and took a long hot shower and shaved. He came back into the room in his bathrobe and stood in front of the filing cabinet where it sat beside the word processor. He stared at it for a long moment. He would never knock himself out again. Ever. When had he made that decision? He did not know, but it was made nonetheless. That was that. He went over and touched the filing cabinet.

"Sit there and rest easy, you little son of a bitch. You've got all of me you'll ever get."

He went into the kitchen and made bacon, soft scrambled eggs, and toast. He shoved aside the remains of the tripe and tongue and sat down to eat, chewing slowly, wondering about the night before and the days to come. And it came as a small shock to him that he felt equal to whatever the days to come might bring. He felt strong, strong in the way he used to feel in his early days in the ring when he faced an opponent feeling, knowing, he would not be beaten. Somehow he would work this shit out and he would not be afraid. He knew, even though he had never admitted it to himself, that he had been afraid—lived with a fear that was like a naked nerve—ever since he'd looked into the mirror in the dressing room after his last fight, his last defeat, and had seen himself crying. And for the first time, he knew, too, that all the days since then he had suffered one defeat after another. He was sick of defeat, sick with defeat. Well, he was not going to let the bastards beat him. He would not be whipped again.

First things first. He got up and cleaned off the table and rinsed the dishes and put them in the washer. He wiped down everything in the kitchen, including the stove, and then went to sit at his desk. He wrote a long, quick letter to his daddy and told him he was thinking of giving up the ring, that he probably would get out now while he still had his senses about him. He told him that fighters can and do take some terrible beatings even though they bring home the prize money. Just because you bring home the money doesn't mean you didn't get a beating. And he hoped his daddy would try not to worry about it but he had been taking some terrible beatings lately, even though he brought home the money. Well, there were other things to do in the world and he would find one to do. He would continue to help them as much as he could, but no matter what happened and what came out of it, he loved them and thought of them often.

He licked the envelope, sealed the letter inside, and stamped it. He had not felt so good in a very long time as he did when he finished the letter. He could feel his confidence growing in just the way it had grown as he had fought his way through thirteen bouts, beating one fighter after another. And with each win the belief, the certainty, grew that he would win again. And he would win again. Yes, but at what? Damned if he knew. Go take his old job back at Jacques'? He would do that if he had to. And if he could not work his way out of the place, if that was to be his life, then that was his life, and he would live it. The prospect did not depress him. He pushed himself back from the desk, laced his fingers behind his head, breathed deeply, and stretched and relaxed.

The thing to do was move, to act, to . . . His gaze fell upon the clothes that Charity had bought him for him to defeat himself for the entertainment of college kids. He sat very still, thinking. Well, if he could get whipped in them maybe he could win in them. He smiled as he thought about going back among the wolves, because that was what he intended to do. But they would not chew his ass this time. If there was going to be any chewing done, he would do it. He remembered what his mama had told him throughout his childhood: "It's no shame in gitten dirty, it's only shame in staying dirty." Now he knew why it had caused him so much pain when Jake had said: "You're one of us." Deny it as he would to himself, he had been one of them. Had been. Not anymore. He did not hate, or even dislike, them for what they were, what they did, how they lived. What they did was their business. For himself, he was going to make a different bed to lie in, and now that he knew dead-solid-certain not so much what he was going to do, but what he would never do again, he might as well look down every road open to him. A man never knew what he might find down one of them.

He got up from the chair, shucked out of his robe, and dressed in the clothes and put on the snakeskin boots. He managed to knot the tie, even though it took him three tries to get it the way he wanted it. He had spent every Sunday morning of his boyhood watching his daddy knot his church-going and funeral-going black tie in the mirror over the dresser in the bedroom. He could have managed the tie that night when he went to the party, but he had not wanted to fumble in front of Charity. He checked himself in the full-length mirror.

"You're taped and laced, and got your cup on, son," he said to himself. "Let's go kick the shit out of somebody."

He called a cab and went out to stand on the curb. It was there in only a few minutes, but as he slipped into the backseat, he remembered that he had thrown away the address.

"The J. Alfred Blasingame building in the CBD."

"Yes sir," the driver said and roared away from the curb. No cabdriver had ever said "yes sir" to him in his life. It was the clothes he was saying "yes sir" to. Strange fucking world.

The receptionist on the thirteenth floor was polite, but she said it was impossible to see Mr. Blasingame without an appointment.

"He is in conference and cannot be disturbed."

"Tell him I'm here."

"I've checked the appointment book, Mr. Biggs. I have my instructions."

"Tell him I'm here." His voice was calm and even, and sounded strange to his own ears.

The secretary looked confused, then worried, a little crease appearing between her eyebrows. "It cannot be done."

"Look lady," said Eugene, in the same calm, flat voice, "I don't want to be ugly here. You say you have your instructions, and I'm telling you you won't have your job if he finds out I've been here and he was not told about it."

She hesitated a moment, lifted the phone, and pressed a button. "There is a Mr. Biggs here and he says . . . Oh, I see. Certainly."

When she put the phone down, Eugene said, "Will he see me?"

She was smiling now, the crease of worry gone. "I was speaking to Ms. Tomal, Mr. Blasingame's executive secretary. And I'm very sorry for my confusion in the matter. It is through that door just there to your right. Please go in. And I'm sorry."

"Happens in the best of families. Don't worry about it."

He turned the knob and pushed through the heavy oaken door. Sitting at a desk was a trim, middle-aged woman who looked like nothing so much as a schoolteacher except for her carefully done hair held in place by a gold comb and her clothing, obviously too expensive for any teacher ever to afford. She stood up and came to him across the deep carpet.

"It will just be a moment." She extended her hand. Her grip

was firm and sure. With her other hand she directed Eugene toward a deep leather chair. "Mr. Blasingame needs just a minute and you may go right in."

Eugene refused the chair, saying, "Thanks, I'll stand. Too early in the morning to sit." He really wanted to stand so he could keep looking at the picture on the pale blue wall above the chair. He recognized it for what it was, a real honest-to-God painting, not a picture of a painting. He had never seen a real one before. He was fascinated by the ridges of paint on the canvas as though it had been put on not with a brush but with a flat stick or maybe a knife, and just as fascinated with what was in the picture: a huge pocket watch in the foreground with a rose growing out of it, and behind the watch a mountain of skulls with a young girl sitting naked on top of it holding a snake in her mouth.

"A striking work, isn't it, Mr. Biggs?"

"Jesus," he said.

"It takes a bit of getting used to, I'll admit."

"I don't know how you work all day with that thing up there."

"But you haven't taken your eyes off it, have you?"

He turned to look at her, and she was smiling. It was a warm, open smile, neither mocking nor condescending, but rather encouraging, just as a kind teacher might have smiled.

"You're right about that, but I still wouldn't want to be in the same room with it all day."

"Perhaps," she said. "But permit me to say that I rather think that if you were in the room with it all day, you would miss it dreadfully if you shouldn't see it ever again. As I say, it takes a bit of getting used to."

"I really don't know a thing about stuff like that."

"Mr. Blasingame collects with a passion. Just as he does everything else in his life. A passionate, learned man with enormous energy." Something buzzed on her desk. She picked up a phone. "Of course, sir. Right away." Again she gave Eugene the same friendly smile. "Mr. Blasingame will see you now. It was a pleasure chatting with you."

Eugene didn't know what to say so he said nothing. As he approached the door behind the secretary, it opened and there stood a tall, thin man in a severely cut black suit. It was Oyster Boy only it was not Oyster Boy. No, not Oyster Boy at all. It was not just that he seemed so tall and strong in the way he stood there in the doorway or that the skin of his face and naked skull was moist and high with color, suggesting great good health and vitality, or that he wore on the bone-thin finger of his right hand a heavy gold ring of some simple design, a ring of the sort that only a man who was very sure of himself could get away with, no, it was something else that Eugene could not say but he certainly could feel.

It quickened his blood and brought the taste of copper to his tongue. And yet, while Mr. Blasingame was solemn, even grave, as he extended his hand, Eugene could see nothing in his expression that was threatening. Then why did he feel threatened? He took a deep breath and told himself to try to remember that this was a guy that crawled around with a dog collar on his neck, led on a leash by a fat boy named Purvis. He was just another human being. Like the rest of us.

When Eugene took his hand, for some reason he expected it to be cold and weak. It was not. If anything, the flesh was unnaturally warm, almost hot, and the grip was that of a young athlete.

"Good of you to come, Mr. Biggs."

"I heard you wanted to see me."

"Yes, yes. That I do. Come in, come in."

Mr. Blasingame turned and strode across to a desk that seemed larger than it was because it had nothing on it but a telephone with rank after rank of buttons beneath it, a small sheaf of papers directly in front of where he sat, and to his left, a gold pen set. Directly behind the desk was a wide ceiling-to-floor window that looked out over New Orleans, and a dozen or more paintings hung on the walls, but Eugene only glanced at them long enough to know that they were there, because now that he was face to face with Oyster Boy who was not Oyster Boy, he could not even remember his fascination with the painting outside in the other office. Oyster Boy J. Alfred Blasingame was fascination enough. Eugene did not know what sort of man he expected to find here, but whatever it was, this was not the man.

Mr. Blasingame indicated a heavy leather chair not unlike the one Ms. Tomal had offered Eugene, and said, "Please."

Eugene was relieved to sink into it, relieved to have something to press his back against and to have the soft, heavy leather embrace him. He saw now that the high color and moist skin of Mr. Blasingame's face and skull were some kind of cosmetic, skillfully applied but cosmetic nonetheless. But if his color was faked, Eugene knew that the rest of him was not: his manner, his strength, and particularly his voice, which was not the voice he had heard in the mansion on St. Charles Avenue. Mr. Blasingame pushed back the sleeve of his coat to look at a gold watch as thin as a half-dollar.

"I'm afraid we do not have much time for this initial meeting. After all, I did not know you were coming. Why did you come, Mr. Biggs?"

"I'd just as soon you called me Eugene. And I came because I heard you wanted to see me."

"We've already established that, Eugene. But I would have given odds that you would not have come here today or any other day if I were a betting man, and I am a betting man. I am also nothing if not a judge of character, or the lack of it. So I don't understand. And I like to understand."

"I couldn't explain it to you if I wanted to. Just say things changed up on me, that's all."

"Ah yes, now, that I understand. It is an uncertain world at best."

"I don't know why you wanted to see me, but I'll tell you out front that what you saw me do at your party is gone forever. I don't do that no more." Then he remembered what he had read in Charity's notes and made himself say it: "I don't knock myself out."

"A wise decision, I should think. I'll come directly to the point. I would like for you to join my organization."

"And what is that? Your organization."

Mr. Blasingame smiled, showing long, wide teeth the color of ash. "I told you, Eugene, that we did not have long to talk today, and it would bore you anyway. Simply said: I own things."

Eugene felt a little stab of anger. "I ain't for sale."

"My, my, Eugene, you mustn't be petulant and melodramatic. It helps neither of us. What I want to do is own a fighter and have you manage him."

Eugene got up from his chair and took a step toward the door before he half turned and said: "I thought I might find something here. That's fucking stupid."

"Sit down, Eugene. Don't waste my time. It costs nothing to listen."

Eugene sat again but said: "Did you git you goddam name on this building coming up with shit like that?"

Mr. Blasingame said: "I'd prefer you not curse in my office." His expression did not change and his voice was matter-of-fact. "I got my name on this building doing a great many things. The fact is that my name is on the building, and that being the case, I can indulge myself in certain . . . things. I can explain it all to you, but unfortunately not now. I'm taking a few people out tomorrow afternoon. Could you come?"

"Out? Out where?"

"I'm sorry. It's this press of time. I have a boat on Paris Road in Chalmette. We'll go downriver to the Gulf. A nice relaxed afternoon. God knows I need it. You'll enjoy it. We'll talk fighting and fighters. We'll talk about you joining the organization. What have you to lose?"

"This ain't some kind of kink, is it? I already saw that movie."

"This is precisely what I said it is. If it were otherwise I would tell you. I am a direct man. An afternoon, perhaps evening, on the boat. Casual dress. There will be about thirty people."

"I don't know, I got a feeling about this. It don't make sense. But what the hell, a boat ride can't hurt. I guess. Never had one before. And something else. If this managing a fighter thing ever happened, which it ain't, it's a guy who'd have to come with me. I want him there tomorrow too if we gone talk fighting and fighters."

"Bring whomever you like. And as many as you like." He smiled his ashen smile. "Within reason, of course. Say up to six people. Good people. Good times need good people."

"You mean that?"

"I never say anything I do not mean. And now I fear we are out of time." He pressed a button on his desk and said: "Ms. Tomal, give Mr. Biggs directions to the yacht and any other assistance he may need. And send in the Allied representatives." He stood up and Eugene stood with him. "We depart at twelve sharp. It would be nice if you and your party could be aboard by eleven-fifteen, eleven-thirty latest."

"This is the strangest thing I ever heard of, but I'll be there."

"I know you'll be there, and I also know, Eugene, that it is not the strangest thing you have ever heard."

"You right, Mr. Blasingame, it ain't even close."

Chapter Nine

They were sailing along through the late morning air in Charity's Cadillac with the top down, Tulip and Charity in the back and Pete on the seat beside Eugene as he drove. The sky was solid blue and without clouds of any sort, unusual for New Orleans. The heat was already oppressive but they did not feel it in the stream of air flowing over them. From time to time Eugene glanced into the rearview mirror and watched Charity as she chatted and laughed with Tulip. But for the color of their skins, they might have been sisters on an outing of some sort, perhaps going to a picnic or a crawfish festival. When Eugene looked over at Pete, Pete gave him a thumbs-up with his left fist and grinned like the end of the world. Pete was wearing white duck trousers, canvas boating shoes, a short-sleeved pale yellow shirt, and a white cap with a gold anchor braided on the black visor.

Tulip and Charity wore similar outfits, slacks and blouses with scarves knotted about their necks. Tulip's clothes were tight and hot red in color, Charity's were soft blue and only suggested the outlines of her body. Both of them, and Eugene too, wore the same kind of shoes Pete was wearing.

When Charity had first seen Pete this morning, she had said: "You look like an admiral in the Brazilian navy."

Pete had showed her the same smile he was showing now, had shown all morning, and said: "Girl, I feel like a admiral in the Brazilian navy. I'm not sure I ain't one."

She had been looking at the cap when she said it, and despite the smile and good-natured bantering of her voice, Eugene knew she did not approve. It had been she who had instructed them on what to wear, but she had not said anything about caps, except to Eugene, saying, "And for God's sake, no yachting caps." But he had forgot to relay that information on to Pete and he was glad he had. He thought Pete looked fine in the cap and wished he had one. He wished a lot of things, including that Charity was not along, but she was, all of them headed toward Chalmette, a city separate and distinct from New Orleans but still a kind of suburb.

Eugene had gone straight to Pete's from Mr. Blasingame's office. Pete let him in and he could see that he and Tulip were sitting down to breakfast at a little table in the tiny living room, a table that had not been there before. There was much in the apartment that had not been there before: a stuffed chair, used but clean with a white doily on its back, where the beanbag had been, bright yellow half-curtains at the window, a small houseplant of some sort that sat in one corner. But most of all there was a smell, or the lack of one that had been there before. The musky, dead air had been replaced by a fresh if not particularly cool breeze blowing through the apartment, and Eugene knew that Tulip had somehow managed to remove the blanket from the bedroom window and open all the other windows. Everywhere he looked he saw the hand of a woman, even to a white bird sewn onto black velvet hanging in a frame on the wall.

"Your timing's bad, Knockout," said Pete. "Would have cooked you breakfast you been earlier."

"It's noon, man. Noon ain't breakfast."

"Noon is my breakfast. Grab some coffee and come set down."

"I'll pass on the coffee. But I got a thing to run down to you."

There was another chair at the little table and Pete gestured with his fork. "I figured you did. Pull up a seat. We downtown in this house now, chairs and everything."

"Place looks great and I don't have to ask or wonder why. How you doing, Tulip? You slowly getting the hair scraped off the belly of this beast? Damn place looks great."

She had been looking from one to the other of them as they talked, smiling as she did, and when Pete would look toward her, their eyes would hold for the briefest of moments, and Eugene, watching them, reminded himself to tell Pete what a lucky son of a bitch he was.

"That's nice of you to say," said Tulip. "Thank you. But I just did a few little things, nothing hardly."

"So lay it down, Knocker. What you got to say here at the crack of dawn."

"For starters, it's Eugene from now on. Forever."

"I told you. Just as soon as . . ."

"I have. It's over. Finished."

"You shittin me."

"I ever told you a lie?"

"None I caught you in."

"Then it's Eugene, champ, until I make what I just said a lie, which I won't."

"How did this happen, man? I'm glad, but how?"

"Too long to tell, and it ain't why I'm here. Will Rankin Shortly pull your shift tomorrow? He'd be paid for it, of course."

"Rank will do what I tell him. He owes me. I pulled his shift more than once when he couldn't get out of the bottle. But what's this with the pay? Who pay? You? I thought you said . . ."

"I did and I have. This thing I've got—but it's too complicated to explain—and I don't half understand it myself, but I don't know, it smells good, Pete. It stinks, too, but it smells good."

Pete forked a piece of toast soaked in egg yolk into his mouth. "You got a great nose"—he chewed, swallowed, and smiled—"Eugene, but it only tracks shit."

"Pete!" said Tulip and slapped him on the shoulder with the flat of her hand.

"Look," said Eugene, "I'm just thrashing around trying to find a way to make a nickel. The minute I quit taking myself out to get other people off, my well went dry. You dig? Ain't no water at the bottom of my well anymore. It's a desert down there. I'm trying to find a way to keep from going back to carrying trays at Jacques'. Maybe I found a way, maybe I ain't. There's a guy named J. Alfred Blasingame. Ever heard of him?"

"No."

"He's got more money than God. Owns a building down in the CBD. I was in his office this morning down there and . . ."

"How would you know anybody in the CBD?"

"You know how I met him, man."

"Awwww, Eugene, shit!"

Eugene rushed on. "I'm invited to go out on his yacht tomorrow afternoon. Don't look like that, goddammit. And don't waste my time, I got things to do. The yacht is all straight shit. Thirty other people are going to be there. A few drinks, a little food, whatever the fuck people do on yachts. And I want you to come with me. I need you to come with me. And bring

Tulip." He looked at Tulip. "Wouldn't you like that, Tulip? Could you swing getting off and coming with us?"

"I never been on a yacht," she said in a small, sad voice. And Eugene knew instantly where the sadness came from. She now knew that he knew she worked at Maxie's doing the thing with the teddy bear, that he had probably seen her work, and she was ashamed. Well, he could not get it back—what he'd said—and she would just have to live with it.

Eugene said: "Darling, I never even seen a yacht or if I did I didn't know what it was. You don't know me as good as Pete does, but he knows I don't shoot people down, never friends. Trust me, we'll have a wonderful time and maybe I can stay out of Jacques' and Pete can get the hell out of the Flesh and Flash. What's a contender, a man who might have won the world, doing running a projector for people from Paducah to get their nuts off?"

Pete had stopped eating and had been staring at Eugene as he talked. His face was calm but there was a hardness to it. "You my cut buddy, Eugene. We been cut buddies for a long time." He looked down at his plate, pushed the remains of the food around, and then looked up at Eugene again. "But I already know more than I want to know about the world you live in."

"You don't know a fucking thing about the world I live in," Eugene said. "You just think you do."

"Yeah. Probably. But still more than I want to. More than I need to."

Eugene sighed and leaned back in his chair. "Blasingame wants to talk fighters. That's it. That's all."

"Say what?"

"He wants to own a fighter."

"I retired."

"Not you, jackleg. He buys the fighter. We—you and me— we train and manage him."

158

Pete smiled. "My, my, I wish I had a pound of whatever you been eating."

"I got an address in Chalmette," said Eugene, "a place where they park them boats. We got to be there tomorrow eleven-fifteen, eleven-thirty at the latest. I'll tell you what Blasingame told me. What you got to lose? Talk won't kill you. And a ride on a fancy boat and a little fancy food won't either."

Pete pushed at his food some more with his fork, his face no longer hard, just concerned. "He know you bringing me?"

"I didn't mention your name. I wouldn't have done that until I talked to you. I told him I was bringing a guy I needed with me. He said bring anybody I wanted to, as many as I wanted to. Up to six people, he said. What can go wrong, Pete?" He smiled. "Anything we don't like comes down we put our backs together and kick the shit out of everybody and take the fucking boat to Key West."

Tulip put her hand on Pete's arm. "Let's do it. It could be fun." She looked straight at Eugene. "I need a break from the Quarter and that teddy bear. I could miss it real easy for a night."

"All three of us," said Eugene, "could miss what we doing for the rest of our lives real easy. This may be a way to do it and it may not. We won't know till we go see. What about it?"

Pete sat very still for a moment and then he gave his wide, fine smile, and put his clenched fists out over the table. "Let's get it on."

Eugene clenched his own fists and tapped Pete's, first on the top and then on the bottom. "We'll give the folks a good show. We come to fight."

"That we did, Eugene," said Pete. "That we did."

Eugene pushed his chair back and stood up. "I got things to do. I'll be in touch." At the door, he turned to look back. "Blasingame did say it was casual dress."

"I got you covered, man. Levi's and a T-shirt, right?"

"Sounds casual to me. Tulip, take care of my man. And yourself, darling. Later on."

Behind the wheel of the Cadillac, a tape of Hendrix turned full volume, Eugene looked at Pete in his fine duck trousers and thought: Levi's and a T-shirt, my ass.

That was what Charity had said when she inadvertently found out about the yachting trip: "Levi's and a T-shirt, my ass!"

She would not have found out about the trip if Tulip had not called. He tried to get to the phone ahead of Charity, but he did not make it in time because he didn't want to seem to hurry. But he knew in his heart that it was Pete and that he would let word of the trip out because he had not told him that he had no intention of taking Charity. Pete didn't know anything about Charity, really, except that she was beautiful and rich. That was the source of his envy of Eugene's being with her.

"Josephine?" said Charity into the phone. And all Eugene could do was stand watching across the room. Charity listened a moment without speaking. "Oh, yes, of course. Tulip. I've heard so many wonderful things about you." Again, a silence. "Well, you thank Pete for me, he's sweet to say that. I only wish I saw more of him, and that you and I could have a nice long chat." Now, as she listened, she cut her eyes toward Eugene. "Well, of course, we will. At long last. I'm glad you called and I think you're absolutely right. Eugene went out but he should be back soon and I'll talk to him and call you. At Pete's? Surely. I look forward to it. Yes. Bye."

She slowly put the telephone down. Her face, even as he watched, became sadder and sadder until she seemed on the verge of tears. He told himself that she was, and had always been, a great actress and that all he was watching was a performance. But the sadness certainly looked genuine.

"What have I ever done for you to treat me like this?" Her voice was as sad as her face.

He started to tell her about having the lock picked, to tell her everything, to tell her that he knew everything. But that would not help anything. And truthfully, he did not know what she would do, was even afraid of what she might do. He was not interested in fighting with her or even in hurting her. Her head was messed up bad and he knew how painful it was to have the kind of head she was living with every day. Besides that, she had carried him all this long time, helped him, and allowed him to help his family.

She sat on the edge of the bed. "I'm terribly hurt. I don't understand how you could invite Pete and Tulip or Josephine or whatever her name really is to go out on a boat—a yacht, she said—and not invite me. And I was embarrassed, for God's sake. She thought I was going and I didn't even know where I was supposed to be going and . . ." Her voice choked off and tears were running on her cheeks and Eugene knew that she was not that good an actress. She was genuinely hurt. She dabbed her eyes. "Do you want to tell me what I've done?"

"Nothing. You done nothing."

"Then why hurt me this way? Humiliate me?"

"I didn't do it to hurt you, Charity. Everywhere I ever took you, you either hated it or been bored shitless."

"But a goddam yacht? Downriver to the Gulf?"

"I've never been on one, or even seen one as far as I know. How was I to know what to expect? Far as that goes, I still don't. You got your work, I don't know, I just thought you'd be busy and wouldn't want to go."

"Well, I'm not that busy, and if you wanted to know about yachts you should have asked me. The only reason Daddy hasn't got one is he doesn't have time for it, but his friends do. I could have helped you."

"Then I guess I made a mistake. It damn sure ain't the first one I ever made." This would probably hopelessly fuck up something down the line, but he had to say it. "Do you want to come?"

"Yes." She smiled and brushed her fingers across her cheeks where the tears had run. "And I'm sorry I acted out that way. I was just so . . . hurt." She picked up a mirror from the side table to check her makeup. Then: "Is Tulip a stage name?"

"I don't know. But she likes to be called Josephine. That's her name, Josephine Apolarus."

In a voice almost as though she were speaking to herself, she said: "I don't blame her for wanting to be called Josephine. The name Tulip must have all manner of associations with . . ."

"Leave it alone, Charity."

She looked up from the mirror where she had been doing something with a little brush to her cheeks. "I'm sorry. Any sort of behavior impinges upon my profession, on my view of the world."

"Just leave it alone," he said. "And try not to talk like that tomorrow. I'll get a headache."

She laughed and came to embrace him. "You darling. I'm very happy. I'm going to meet Tulip and see Pete again whom I happen to admire enormously, and thank you for inviting me."

"Try to leave your job at home tomorrow. Probably be good for you. You think you can do that?"

"Of course I can do that. And I will. We're going to have a wonderful ride down the river. A few drinks, a few laughs. I'll leave everything else at home, and you're right. It probably will be good for me."

"You want to tell me what the call was about?"

"Tulip—Josephine—was concerned about what to wear, not just for herself, but for Pete too. She said something about Levi's and a T-shirt."

"So? I told him casual, Pete said Levi's and a T-shirt. Sounds good to me."

"Levi's and a T-shirt, my ass! These people don't know us. If they did, and we were rich enough, we could show up in sackcloth. But since that's not the case, we'll dress accordingly.

162

Leave it to me. Give me Pete's number and let me call Tu . . . Josephine back."

Eugene turned the sound down on the music as he wheeled onto Paris Road and half turned his head against the roar of the wind. "All right, children, look alive. Cajun Haven is the place we looking for. It ought to be somewhere down here on the left. Forty-two twenty is the number, but I don't imagine we'll see a number."

"Man," said Pete, as they passed a marina. "Look at all them boats. I never told you but I can't swim. Not a lick. Go down like a fucking stone."

"Be cool," Eugene said. He winked at Pete. "Drowning's a easy death, I been told, a real easy death."

"Thanks," said Pete, but his great smile had never left him, and Eugene felt good, felt like he could handle whatever was waiting for him.

Out of the corner of his eye, he saw Pete adjust his cap, setting it at a more rakish angle. The cap was the only thing Pete was wearing that he had bought himself. Charity had talked quietly on the phone with Josephine for a long while and then left in her Cadillac. She didn't say where she was going, and he didn't ask. Where she went was shopping with Josephine. Josephine did not have to show up at Maxie's in the Quarter until nine o'clock in the evening and they had time not only to shop but for an early dinner together, too. When she got back with the clothes she had bought and told Eugene about the two of them having dinner and about how well, as she put it, they fit together, he could hardly believe it, and she could see it in his face.

"Don't look like that," she said. "I get along with anybody and everybody because—in a sense—I am anybody and everybody. To put it the way they would say it on the street, my bag is other people's bags. She's a delightful girl. You neglected to mention the fact of how deeply she loves Pete."

"That's new," he said. "She didn't always."

"Nothing is always," she said. "Try these things on." Everything fit, but he had known it would. "Josephine had Pete's sizes by rote as if they had been married fifty years."

Later that night, lying in bed, she kissed him and touched him and her hand moved, almost imperceptibly at first, toward the Dictaphone.

"No," he said.

Her hand stopped, then moved to his hair and stroked it. "I understand, darling, if you don't feel like telling Charity now."

Not ever, he thought.

She lay very still, stroking his hair. Finally, she reached for the light. He half turned and her hands moved slowly over him in the darkness but finally stopped, the fingers of one hand caught gently in his hair.

"Was this yacht trip the result of last night?"

"Last night? In a way."

"Will any of your other friends be there?"

"Friends? What friends would that be?"

"I mean people you know through . . . by the . . . from the gigs."

Very slowly, a little space of silence between the words, he said: "From knocking . . . myself . . . out?"

She took her hand out of his hair. He heard a truck pass on Magazine Street. And then finally: "Yes. From that."

"I don't think so. Maybe." Blasingame did not count. Charity would never find out he was Oyster Boy. But Jake might be there with or without Femally. He counted Jake as his friend. She was good people, and the last thing in the world he wanted was to have Charity meet her. But if anybody could take care of herself, surely Jake could.

"I'm looking forward to tomorrow," she said. "I wish it was already here."

"I just hope you understand this ain't kink. This ain't showtime."

"I understand," she said.

They all saw the sign at about the same time, but they would not have if they had not been looking. It was only about four feet square with black block letters across a ground of blue that read: CAJUN HAVEN. Underneath the words some sort of fish was leaping free of the water.

"Don't seem to be advertising much for new business," said Eugene.

"Where would they put it?"

"Good point, Pete," he said, "good point."

Eugene estimated there were three hundred acres of boats tied in slips behind the gate where the sign hung from chains, at the main entrance. He had no trouble finding a parking place in the huge lot. He put the top up and secured it against the unpredictable Louisiana weather, although the sky was still a solid, blazing blue. The Cadillac looked right at home amongst the Porsches and BMW's and Mercedeses.

They stood by the car a moment in silence and then Pete said: "That's some heavy gravy you looking at out there on the water, Jack."

"I told you," said Eugene. "This guy's got more money than God. But you ain't wrong. That's serious cheese."

Tulip stood closer to Pete and took his arm, her beautiful little mouth slightly open, breathing through it as though from sudden exertion. Only Charity took it all as a matter of course, or so it seemed to Eugene. She shook her hair out and stretched.

"What a beautiful day to be on the water," she said. "Don't you think, Jo?"

Eugene looked at her. It was Jo now, was it?

"What now, chief?" said Pete.

"We looking for dock F, all the way to the end, the *Agincourt*. That's the name of the boat."

"Let me see," said Charity, looking at the little slip of paper

Eugene was holding. "The way you say that, Eugene, is 'Ah-zhin-coor.' Either say it that way or don't say it once we get on the yacht."

"When they start buying my grits, they can tell me how to talk."

"I'm only trying to be helpful, darling."

Tulip touched Eugene's arm. "Ah-zhin-coor," she said. "She's only trying to help us not look like fools in front of people who probably think they're better than we are anyway. Me? I'm used to scamming the marks."

"Ah-zhin-coor," said Pete. "Sounds good to me."

"It was a battle between the English and French a long time ago," said Charity. "The English won and the English king, Henry the Fifth—twenty-seven years old at the time—ordered the slaughter of the French prisoners."

They were strolling down a broad cement walk off which the docks were at right angles—four abreast, two couples holding hands—and anyone looking at them would never have guessed at the rage in Eugene's heart. He was tense, edgy, trying to psych himself into a game face. He wanted to walk away from this Saturday afternoon with something better than he had, and he did not need any trouble from Charity.

"I thought you said you'd leave the bullshit at home today," he said bitterly.

Pete said: "Hey man, get with the program, lighten up. That's interesting stuff. Henry musta been a badass at Ah-zhin-coor. Go on, try it, Eugene. I bet your Georgia mouth can't even fit around it."

Eugene smiled at him and was grateful Pete was with him today. "Ah-zhin-coor."

"Now we got everything right," Pete said.

"I didn't think I said anything wrong," Charity said softly.

"I'm sorry," said Eugene, and he was. He got no pleasure from hurting cripples, and that was how he thought of Charity.

Besides being crazy and dangerous, she was a cripple. She didn't ask for the head she was carrying any more than he had asked for his. You don't have to like your head, you just have to live with it.

They found dock F easily enough, and the men, women, and children passing them on the cement walk, carrying ice chests and folding chairs and fishing equipment and baskets of one kind or another, became more numerous. Pete and Tulip fell behind Charity and Eugene to make it easier to walk among the stuff piled on the dock at each slip where boats were tied. The people they saw were dressed in every conceivable fashion, some as the four of them were dressed, but some naked to the waist in denims, some in khaki shorts, and many in bathing suits.

"Some of these boats are bigger than the house I grew up in," said Eugene.

"All of them are bigger than the one I grew up in," said Tulip.

"Nice boats," said Charity. "You'd be stretching a point, though, to call them yachts."

"How bout that down there, sweetheart? Will it do?" said Eugene.

"It will indeed," she said, "in every way."

"Sweet mother," said Pete. "That's a fucking ship. It can't be the one we looking for."

"It's at the end of the dock, man."

"There," said Charity, "toward the fantail—the back of the boat—look there. *Agincourt*."

"Ah-zhin-fucking-coor," said Pete.

"Pete," Eugene said, "you and me going to be the only two niggers on this ride."

"Tulip makes three," said Pete.

"Color don't have nothing to do with being a nigger," said Eugene. "I thought you knew that. It's only two. You and me."

"You the two strangest boys I ever known," said Tulip.

Eugene said: "You don't really know us yet. After you know us, you won't want to know us."

"How long you think?" said Pete.

"Two hundred feet?"

"Something between one and a quarter and one and a half," Charity said. "At a cost of about three thousand dollars a foot, give or take, depending on the equipment aboard and how it is appointed, it's quite a plaything."

The craft was solid white except for red and black bands trimming the hull, and women and men wandered over the deck holding drinks or stood in couples or small groups at the rail talking. A New Orleans rag, a strong horn with a guitar and drums behind it, drifted out from somewhere. A railed gangplank led from the dock up to the yacht. Several people stood at the foot of the gangplank, laughing and drinking. One young man, tall and lean and very blond, a little apart from the others, moved alone to the music. He was wearing bleached-out and wrinkled denims and a tuxedo jacket. Eugene pointed him out to Charity.

"Either rich, eccentric, or not invited," she said. "One, two, or all three."

Eugene stopped a few feet from the gangplank and looked at Pete. "You ready to do it, bro?"

"Settin dead on ready. Ain't, I won't ever be."

"Oh, you guys, don't be so serious," said Tulip. "This is sposed to be fun, remember."

"You heard the lady," said Charity, "fun and good times is the order of the day."

Eugene led the way up the gangplank, where a young man in a severely cut, short white jacket over dark blue trousers waited holding a clipboard.

"May I have the gentleman's name and party number?" the young man said. He had a slight Cuban accent.

Eugene only stared at him.

"Eugene Biggs and a party of three," said Charity.

The young man consulted his clipboard and looked up smiling. "Very good, madam, thank you." He turned to Eugene. "Mr. Blasingame requests the pleasure of your company on the afterdeck upon boarding."

To Charity Eugene said: "I could have handled the name; it was the party shit that threw me."

"Ahhh, Eugene, my man, you doing this up big time," said Pete.

On board, the guests were dressed as those at the foot of the gangplank had been, in everything from bikinis to blazers over white trousers to clothes Eugene would have called seedy. He was in fact admiring a bikini—or rather the body of the girl in it—when he realized the girl was Jake. She had the perfect body for the bit of cloth she was wearing, thin enough to appear voluptuous when she was nearly naked. He saw her at the same time she saw him. She turned from a young lady she was talking to and waved.

"Friend of yours?" said Pete.

"Yes," said Eugene.

"Nice friend," Pete said, "very nice."

Tulip smiled. "You mine, sucker."

"Everything but my eyes, Tulip," said Pete. "They belong to everybody. I didn't go blind when I got tight with you."

"Don't you think you might introduce us?" Charity said.

It looked as if he was going to have to even though he wished he didn't, if for no other reason than what he had told Charity about Jake. He genuinely liked Jake and he felt—knew—he had betrayed her trust. They walked down the deck toward the rear of the yacht where Jake was leaning with her elbows on the railing.

She showed her incredibly white, sharp little teeth: "The meeting must have gone well," she said.

"Well enough," said Eugene.

"He didn't seem much like Oyster Boy in the CBD, did he?"

she said. Eugene did not so much see as feel Charity stiffen beside him. Jake pointed and when Eugene turned to look, there was Mr. Blasingame some distance away sitting at a table under a huge white umbrella and wearing a wide-brimmed white hat. There was a telephone on the table, and three Japanese men dressed in conservative dark blue suits sat with him, briefcases sitting beside their chairs on the teakwood deck. "And he sure as hell is not Oyster Boy sitting there on the back of this monster boat."

Jake's breath was heavy with the sweet smell of tequila and her eyes were a little glassy with it and her smile a little too fixed and determined. Remembering the night in her apartment, Eugene knew she had to have been working pretty hard at the bottle to be this far along so early in the day.

"I've not met your friends," said Jake.

Charity put out her hand and said, "Eugene is nothing if not informal. My name is Charity Beechum."

Jake took her hand. "Jake Purcell." They remained with their hands gripped firmly, their eyes holding in a steady gaze.

With her hand, Charity turned Jake slightly. "These two delightful people are Pete Turner and Josephine Apolarus. Pete goes back with Eugene all the way to his days as a professional fighter."

"I'm glad to meet all of you," said Jake. "We might even have some fun today. And Eugene, quit looking so grim, for God's sake. This is a piece of cake. Easy enough so far, right?"

"Yeah, easy enough so far," Eugene said. "And thanks, Charity, I ain't used to introducing people. Where I come from, we all know each other."

"Eugene, you're not where you came from now," Jake said.

Charity said: "Precisely what I'm always trying to tell him."

Pete, relaxed and obviously pleased to find himself on a yacht, slapped Eugene on the ass, and said: "Ain't neither one of us where we used to be from, and with any luck at all, we never will be again." He turned to Josephine who had stood quietly through the introductions, her pleasant brown eyes turning to follow the conversation as each of them spoke, and said: "Baby, when you going to quit hogging the conversation? This girl's only bad habit is she'll talk your ears off, but she's young and maybe she'll grow out of it."

Josephine slapped him playfully on the arm, pleased, and at the same time a little ill at ease. "This boy talks a lot of shit but he's cheap to feed."

"The feller that checked us on said Mr. Blasingame wanted to see us," Eugene said. "So I guess we ought to step on back there and talk a minute."

"Jesus," said Charity, "I don't feel like listening to you guys talk about boxing. I need a drink and I need something happy. This is supposed to be a party."

Jake, who had only a moment ago released Charity's hand, took it again. And she took Josephine's, too, saying: "I had a date but it looks like I've been stood up. I'll show the ladies around, load'm up on a few party favorites, and we'll catch you guys later. Talk as long as you like, talk a long time."

Josephine glanced at Pete and he said: "Cool."

As the three of them started off down the deck, Jake already saying something that had made all of them giggle, Eugene called: "And Jake?" She looked back over her shoulder. "Thanks," he said.

"For nothing," she said. "Besides, you've already paid me back."

Eugene guessed Femally didn't like yachts. That, or else she had found something she liked better.

Pete watched them go and said: "It's about half of what just went down that I wish somebody would explain to me."

"No, you don't," said Eugene. "You just think you do."

"Why is it when you say something like that to me I always believe you?"

"Because I told you I never lie."

Pete said: "I don't think I want to chew that anymore. So,

171

my man, unless you want to stand here and chew it until it quits moving, let's go see the man, see what kind of shit he's selling."

"Let's do it and say we did."

Mr. Blasingame saw Eugene coming across the deck and waved. When they got to the table, he stood to receive them. "Eugene, so good of you to come." The three Japanese rose and Mr. Blasingame turned to them. "Enjoy yourselves, gentlemen. And we'll talk more later in the day. Now, if you will excuse me." The Japanese bent for their briefcases, nodded once to Mr. Blasingame and once to Pete and Eugene, and left the table walking one behind the other. Mr. Blasingame said: "Sit down." He offered the chairs at the table with his hand, palm up. "Could I get you something to drink?"

"Not for me," said Eugene. "I'm not much to drink."

"I am," said Pete. "And this looks like margarita weather and a margarita place to me."

Mr. Blasingame smiled and almost imperceptibly moved his hand. A young man was instantly beside the table. Mr. Blasingame ordered the margarita and an iced carrot juice for himself. Then he turned to Pete. "Direct. I like that. You never have to guess about a man who knows and speaks his mind. You were a great joy in the ring, Mr. Turner. The opponent never had to look for you. You were always right there, on his chest."

Pete smiled, but suddenly seemed nervous, even a little embarrassed, and he looked down at his fists, now clenched, as though he expected to find gloves on them. "Well, I'll be dipped in . . ." Eugene had warned him about cursing. "Well, I'll be dipped."

"I was ringside when you took Marvin 'Little Man' Turnhill." Mr. Blasingame closed his eyes briefly and said: "Two minutes of the third, fighting inside, a short, chopping left hook."

"Man, you got me in your hip pocket forever," said Pete, no longer smiling but now serious as death.

The waiter was back with the juice and Pete's drink. Mr. Blasingame took a tiny sip and Pete poured his margarita—all of it—down in a single swallow. The waiter was instantly at the table to take the empty glass.

"Eugene, why didn't you tell me you were bringing the Pistol? The .22 pistol that fired a .45 caliber slug. I'm honored."

"I didn't know you was a fight fan," Eugene said. He had listened to all this with his jaw slightly open. "I didn't know his name would mean anything to you." The waiter had returned with another margarita and Eugene said: "Bring me one of them, if you would. Make it a double."

"For curiosity's sake and nothing more, because I am genuinely honored, how did you happen to bring him?" said Mr. Blasingame.

"He's my best friend."

"First after Charity," said Pete. "She's his main squeeze."

A slightly distasteful look came on Mr. Blasingame's face at the words *main squeeze,* but it passed quickly. "One doesn't like to bring up unhappy memories, but he was your last fight, Eugene, and he took you out hard."

"Hadn't of been him, would of been somebody else. My string was played out. It was over before I ever got in the ring with him. I just didn't know it."

Pete tapped him lightly on the shoulder. "It was a hell of a fight for four rounds, though. You was whipping me like a child till you walked into that right hand, opening seconds of the fifth."

Eugene shrugged. "But I did walk into it. And it did turn out my lights."

"And you're best friends?" said Mr. Blasingame.

"I love him like a brother," said Eugene.

"We are brothers," said Pete.

"Go to the wall with him or for him," Eugene said.

"Remarkable," said Mr. Blasingame.

"But what counts here" said Eugene, "here today, is Pete

has forgot more about boxing, the whole fight game, than I ever knew. You want a fighter? That's fine with us. If we can make a nickel on you wanting a fighter, that's better with us. But first things first. We don't do nothing, until we got a fighter. Everything else is just talk."

"Right to the point," said Mr. Blasingame. "I like that. And getting a fighter does not present a problem."

"Getting the right fighter is always the problem, the biggest problem," said Pete.

"We get a proven fighter and go from there," said Mr. Blasingame.

"If he's proven and on the way up," said Eugene, "he's already locked in so tight you couldn't pry him loose with a crowbar."

"But I'll not use a crowbar." Mr. Blasingame smiled. "I guess both of you have heard of Bobby Barfield, fighting out of Baton Rouge?"

"Friend of ours," said Pete.

"I buy his contract and you manage him."

"The Bad Man would tell you to suck yourself off," said Pete.

Mr. Blasingame looked in genuine pain. "Please. Not that sort of language here."

"But that's what he'd tell you," said Eugene.

"And the guys who own him would tell you worse," Pete said.

"I wish you would leave these things to me. I can get the contract."

"I don't know what all this is to you, owning a fighter," said Pete. "But fighting is the Bad Man's life. It's his blood and bone. Getting his contract don't mean nothing. He wouldn't fight for you."

"We'll see."

"You messing around in the man's life," said Eugene. "You messing bad. Bobby ain't setting on some yacht in the sun. He's in the gym today sweating blood."

174

"I ain't talked to Eugene about this, Mr. Blasingame," Pete said. "But I'm putting it this way. I know where Bobby's been, I know where he is, and I know where he's going. I know about paying dues. I love the Bad Man and I want no part of this. None. You fuck—and that's right, I said fuck—with Bobby and I'm out. Gone."

"What do you say, Eugene?"

"Ain't nothing to say. Pete just said it all."

To their great surprise, Mr. Blasingame laughed and waved the waiter over for another round of drinks. "I'm pleased. I could not possibly be more pleased. Honor is difficult to find in the world and very nearly impossible to keep."

Chapter Ten

With the exception of the time it took Mr. Blasingame to take two calls on his cordless telephone, the three of them talked boxing without interruption all the way downriver to the Gulf of Mexico, and by the time the yacht broke into open water they had the terms of their venture settled to everybody's satisfaction. Eugene was impressed with Pete in a way he had never been impressed before. Eugene would not have known what to ask for from Mr. Blasingame. He had never had any kind of signed agreement with Budd, nothing even verbal that was very specific. But Pete was specific, very specific. About everything.

"This whole thing is going to cost a lot of money," Pete said.

Mr. Blasingame smiled his humorless smile. "Doesn't everything?"

"If the conditions are right, I can find the boy, but even that won't be cheap."

"Anything worthwhile never is," Mr. Blasingame said.

"You ought to know going in that with the breaks going our way—and they probably won't—nobody may see a profit for three, even five years. The only money in any division is for the top two or three guys. The only real money is having a champion, a popular champion."

"Then for five years, perhaps forever," said Mr. Blasingame, "it will simply be a tax write-off, won't it?"

"I wouldn't know about that," said Pete.

"My accountants would. And in the meantime, it would give me pleasure."

"A very expensive pleasure."

"Most of my pleasures are."

This talk of money and accountants and championships had stunned Eugene more than a little. Besides that, Pete sounded as though he were trying to talk Mr. Blasingame out of owning a fighter, talk the two of them out of a job. Eugene glanced at the waiter hovering just out of earshot and the waiter was instantly at the table, where Eugene handed him his empty glass. Pete pushed his own empty glass toward the waiter without ever taking his eyes off Mr. Blasingame or missing a beat in what he was saying. Eugene watched him, amazed. Was this the same guy whose bedroom was hardly bigger than a closet and who had slept in rank, dead air until he paid a hundred dollars to a girl, who did a nasty act with a teddy bear, for a howling mob, to fall in love with him for six hours before she finally fell in love with him forever for nothing?

Mr. Blasingame spread his dead-white bony fingers on the table and said: "So that we may enjoy the afternoon and the other guests, why don't you just put it all in front of me. You too, Eugene. You've been too quiet."

"Pete was a better fighter than me," said Eugene, "and he's doing this better than I could, too."

Mr. Blasingame frowned. "I hate to see a man sell himself short."

Eugene said: "Who's selling who short? I brought the best mouth in New Orleans with me, didn't I? He sound like he knows what he's doing or he sound like he knows what he's doing?"

"A point well taken."

"A really good kid," said Pete, "say a National Golden Gloves champion, will still usually have to hold down a job and train in the evening. But if we can feed him, give him someplace to put his head down, put clothes on his back, a little pocket money on his hip, and cover his gym expenses, we can have the pick of some very good boys. Eugene and I will go straight salary, say four hundred apiece a week, plus expenses. Kick it up a hundred every time the kid wins, lose a hundred every time he loses. You own, Mr. Blasingame, twenty percent of the kid forever. Eugene and me take ten percent apiece. We never make a nickel on our ten until you make back what you pay us and what it costs to keep the boy. The boy knows all this going in. He starts taking sixty percent when his fights make back what we—you—got in him. It's a good deal. A fair deal. We'll get a contract, lawyers, make it right. Nobody gets fu . . . screwed."

"I have no shortage of lawyers," said Mr. Blasingame.

"Fine," said Pete. "Use'm. Eugene and me'll get our own."

Mr. Blasingame said: "If I were not hiring you for this, Pete, I would hire you for something else. You're a good man at the table."

"Then it's a deal?" said Pete.

"No," said Mr. Blasingame.

And so with Pete and Mr. Blasingame bargaining back and forth, Eugene drank steadily as they sailed through the bright air down the Mississippi River. It all got more and more complex and more and more specific. Mr. Blasingame did not want to give up twenty percent of his fighter. Pete settled on fifteen percent—seven and a half percent for himself, the same for Eugene—but in return, Pete got the salary raised to five hundred a week. To go up to five hundred, though, Mr. Blasingame made the bonus in salary for the fighter winning only fifty dollars, while the penalty for losing was still a hundred-dollar reduction in salary. Pete and Eugene had the right to sell their interests in the fighter, but Mr. Blasingame retained first option to buy them out. The right to sell and first refusal option were not, on the other hand, reciprocal. Pete pointed out that neither he nor Eugene had transportation, unless you could call Eugene's motorcycle transportation, and surely Mr. Blasingame did not want his fighter's handlers riding around on a motorcycle. Certainly he did not. He had a lease agreement for automobiles for certain of his employees. Mr. Blasingame offered a Ford Maverick. Pete wanted a Cadillac. Eugene wanted another drink. Through a vodka mist that grew gradually redder, Eugene could see how much Mr. Blasingame was enjoying this bargaining and jockeying for leverage, which did not surprise him at all. After all, the man owned a fucking building in the CBD, and a mansion on St. Charles Avenue, and they were sitting on his yacht. What did surprise him was how much Pete was also enjoying the give-and-take, and how good he was at it. And Pete, to his certain knowledge, did not own a goddam thing, except a competitive instinct and a will to fight that might have won him the world. Eugene, happy and stunned with drink, relaxed and watched the seagulls wheeling and turning over the water. He was so happy and so stunned, in fact, that he did not realize the discussion was over until Mr. Blasingame was already on his feet.

"I look forward to working with you," said Mr. Blasingame. "The finer points are best left until later. I have guests to look after. May I show you about the craft? Introduce you to some of the other guests, perhaps?"

"I think I'd just like to sit here and think about all this for a minute," Eugene said.

Pete said: "We've got a couple of ladies aboard somewhere. We'll run'm down after my man here gets his sea legs."

"They're with Jake," said Eugene.

"Then they're in fine hands," Mr. Blasingame said.

"She's good people," Eugene said. "I didn't always know that, but she's good people." Only when he heard himself say what he said did he realize how over-the-top drunk he was.

"That she is, Eugene," Mr. Blasingame said. "That she is."

"We're going to do it up tight and right," said Pete. "Good doing business with you."

Mr. Blasingame lifted his glass of carrot juice in toast. "Pleasure, gentlemen. Let pleasure rule the day." He turned and strode across the deck.

"Absofuckinglutely amazing," said Pete when he was gone.

"Some better than I thought it might be," Eugene said.

"I'll believe all this when I see the money."

"I don't think it's jive, man. I think it's for real."

"Cut buddy, you ain't said shit that whole time. I'm in up to my ass and don't know half what I'm talking about and all you do is sit here breathing and drinking. Drinking some. Tell me I'm wrong. You are about three-quarters twisted, ain't you?"

"Why should I say anything when I got you," said Eugene. "And yeah, drunk some."

Pete reached over and threw his arm around Eugene's neck. "You sweet son of a bitch, I may shuck Tulip and marry you. This is a bitch of a day, sweetheart. I don't know how you fell in this and I don't want to know, but Jesus, thanks for the leg up for me."

"Why ain't you drunk?" said Eugene.

"Because I was talking. *You* was drinking."

"You think they got any food on this thing? I need a bottom for my stomach."

180

"What they ain't got on this boat, nobody needs. You want food, we find food. Come on."

Eugene put his hand on Pete's arm. "Hold it a second."

"Yeah?"

"About Mr. Blasingame."

"What about him?"

That was exactly what Eugene had been turning over and over. How much of what he knew of Mr. Blasingame should he tell Pete? Anything? Everything?

"He's . . . he's a shark."

"Eugene, you drunk and you do need a bottom. Bad. Where the hell you think I been all my life? I don't know how the dude got his bucks or what kind of tree he swings from, but two minutes after I started talking to him, I saw that old familiar face, man. That face been gnawing my ass since I left my mama's knees. They ain't two cents' worth of difference in that sucker and the jiveass iceman I been making it on the corner with my whole life. Be easy, sweetheart, Pete know these waters. I been here before."

"You ain't been here before."

"You know how much of my ass I owned when I was seven in the world? I never told you this. You know how much?" Eugene only sat watching him. "Not a fucking ounce. I'd been beat out of ever piece of my skin. Everybody else riding and I was walking. It ain't gone happen again. Count on it. Rest easy. Now let's you and me go find that food." He turned and pointed to a huge glass-enclosed room in which they could see men and women standing about with glasses in their hands. "Wouldn't surprise me if it was food right in there somewhere."

"I magine," said Eugene.

"You need a shoulder?"

"Steady as a rock, man," Eugene said, but he put a hand on Pete's shoulder when they stood up.

The waveless, blue-green Gulf was flat as a pond, and walking over the deck was like walking on the street. Pete opened the door and Eugene immediately felt a little better as the chill air of the room washed over him. The four old black men playing muted blues at the far end of the room on a little platform covered in blue velvet reminded Eugene of the college students' party. He had not known what he was walking into then, and he reminded himself that he did not know what he was walking into now.

"Look yonder, old son," said Pete.

A long line of food was set up at right angles to the corner where the musicians, looking bored and sleepy, played. Two men in starched linen jackets and wearing high chefs' caps stood looking equally bored and sleepy because nobody was paying much attention to the food they were obviously there to serve. The action centered on the round mahogany bar in the middle of the room where two bartenders worked to fill and refill the glasses being shoved at them.

"Well, I thought you'd never," a voice said as a soft, heavy hand touched his elbow.

Eugene turned his slowly spinning head, trying to stop the spinning but fearing it was going to spin faster. It was Purvis, enormous in a satin jumpsuit which the creases of his fat caught and held under his sloping breasts, on his stomach, at his groin.

"We'd never what, man?" said Pete pleasantly.

"Come in from that savage, savage heat and join the party."

"I don't believe I know you, biggun," Pete said.

Purvis looked at Eugene. "Aren't you going to introduce me to your friend, Eugene?"

"No."

"That's not the right answer," said Purvis, "and not the right attitude." He put out his thick, soft hand. "My name is Purvis, Purvis Reeker. I'm Mr. Blasingame's man."

"Hell, ain't we all," Pete said, shaking hands with Purvis. "And don't mind Eugene here. He sometimes gets out of step

182

with the parade. But he comes back quick, Eugene does. My name's Pete Turner and . . ."

"I know," said Purvis.

"You do?" said Pete.

"I was just giving Eugene a chance to be polite. You're here with Miss Josephine, perfectly delightful creature, and just there"—he pointed across the room to Charity, who was talking with Mr. Blasingame while holding lightly to Jake's arm—"and just there is Miss Charity, the third member of Mr. Biggs' party."

Pete slapped Purvis good-naturedly on the ass and said, "Now ain't you a hip, magic motherfucker."

Purvis blushed and his fat face jerked in a shy smile. "Not magic at all, really. Ernesto—he's the one who met you when you came aboard—is my man."

"Your man?" said Pete.

"In Mr. Blasingame's world it's like medieval times when every man had his man and God had the king."

"Kiss my ass," said Pete, "God had the king, you say."

"And Ernesto reported immediately to me because he knew I needed to know because I'm Mr. Blasingame's man and since you're one of the special ones on board today, I'm here to make sure you have everything you need."

"Say everything I need?"

"Or want," said Purvis. "Alas, they're not always the same." And again a faint blush rushed up his creased neck and chins into his face.

"Hear that, Eugene? I'm special and I get what I want. Or need. Afuckinglas."

"Oh, everyone in your party aboard is special today."

Eugene said: "Will you believe me if I tell you it's raining, Purvis, when I puke on you shoes?"

"Do be happy," pleaded Purvis in a voice that was sad and strangely girlish, his face now darkly frowning, "and don't say terrible things. You'll just spoil the day."

"We need to feed him," said Pete. "He get cranky when he get hungry."

"Well, for God's sake, for God's sake, I can understand that. Why didn't you say so? Let me take care of that for you." He pointed again. "Take that table back there. I'll have a steward bring you something wild and Cajun and delicious. Everybody else is ignoring the food. I'm glad somebody is hungry. I certainly appreciate that."

He danced away toward the buffet table in a step surprisingly light for a man so heavy. They watched him go and Pete said, "Boy's got his full growth, ain't he? He ain't but a biscuit away from four hundred pounds. How come you so mean to him?"

"Because I don't like him."

"Ain't a hell of a lot you do like anymore, seems like."

"Seems like," said Eugene, raising his hand to take Pete's shoulder again. "Let's go sit down."

They took a table covered by a white linen cloth beside a tinted glass wall through which they could see oceangoing freighters steaming for New Orleans and sailboats—everything from two-man catamarans to three-masters as big as the yacht they were on—and heavy, low-riding shrimp boats with nets winched into their rigging. The water through which they moved was a blue mirror reflecting light.

A waiter appeared at the table almost before they had a chance to settle in their chairs. He carried a round tray on which there were glasses of iced water and hot bread wrapped in cloth in wicker baskets and dishes of mounded butter. "Would the gentlemen like wine with lunch?" he asked.

"I'll have vodka straight up," said Pete. He hooked his thumb at Eugene. "Coffee for him."

"Very good, sir."

"No coffee," said Eugene. "A vodka for me, too."

"Of course, sir," said the waiter and was gone.

"I thought you had the dark twirlies, man."

"I said I wanted a bottom, I didn't say I didn't want a drink. We've done our business, I'm ready to go home. But we're in the middle of the fucking ocean. I'm not going to hold up to the rest of this sober. Need a little grease for the machinery, boss man, a little grease."

"You strange, Eugene, you know that? Hold up to what? The man said we got anything we wanted, so what's the problem?"

"No problem."

"Good."

They watched Purvis coming toward them, driving two waiters carrying trays loaded with dishes before him. As they put the food on the table, Purvis said: "Here's a little bit of heaven for you. My own favorite. I do hope you like it. Roasted quail stuffed with crawfish dressing, a touch of corn maque choux, and a steaming bowl of gumbo." He shuffled where he stood, his great body swaying in a little dance of anticipation. "Well, go on! Go on! Try it!"

Pete ripped a drumstick about the size of his thumb off the quail, put it in his mouth and stripped the meat off the bone. He chewed slowly and swallowed, nothing showing on his face. Then he cracked the little bone with his teeth and sucked the marrow.

"Well?" said Purvis.

Pete looked up from the dish and a great, open smile spread over his face. "Purvis, your taste and my taste fit like spoons. If you was a woman, I'd marry you. Man'd have to die and go to heaven to git better meat than that."

Eugene, who had been sucking a piece of bread he'd torn off a hot loaf and soaked in gumbo, said, "You done good, Purvis, you done real good." He sipped from the ice-cold vodka the waiter had brought in a frosted goblet. "Everything is just right."

Pete had ripped open the quail and was spooning the dressing into his mouth. He paused with the spoon in the air, and looked about the room. "Say, Purvis, where's Tulip?"

"Tulip?" said Purvis.

"Josephine," said Eugene.

"Uh, she's below."

"Below what?" Pete said.

"Decks."

"Break it down for me, Purvis, I ain't a sailor."

"Downstairs. She's downstairs in the greenroom."

"And what would that be, the greenroom?"

"That's where people wait."

Now Pete put his spoon down and pushed his plate aside. "Wait for what?"

"For what they want. See, it's called the greenroom because in the theater . . ."

Pete waved him silent. "I don't need to know nothing about that shit. What's my woman *doing* down there?"

"Anything she wants, I suspect," Purvis said in a voice that was at once shy and contemptuous.

Pete stood up, reached for the front of Purvis's jumpsuit, and jerked him close, their noses almost touching. "Talk straight, man. I don't play when I'm talking about my woman."

Purvis leaned to the side and whispered something that Eugene could not hear. He could not imagine why Pete was so hot, but Eugene was more than a little drunk, and he knew it. Maybe Pete knew something he didn't know.

"Show me," said Pete, releasing Purvis. Then to Eugene, "Hold it down here, Eugene. Back in a flash. Got to check on something." And he was gone, following Purvis through a doorway on the other side of the bar.

Eugene would have gone with him but he did not trust himself on his feet. He finished the food and sat sipping from Pete's drink after he had finished his own. Charity was sitting beside him before he saw her. Jake didn't sit down but lounged there before the table. She had changed her clothes and was now in a white cotton shirt and trousers with a light-yellow snap-brim Panama hat pulled low over her eyes. An unlit black cheroot was caught between her pouting lips.

"Mr. Blasingame says you're going to do business," said Charity.

"We'll see," said Eugene.

Jake cocked one of her thin model's hips and said: "Why so pale and wan, fair lover?"

"What?"

"Why the hangdog glum?" she said. Then: "What the fuck ails you?"

"I don't think I like boats," he said. "And these ain't my kind of people."

"Well, I want you to know you've made me very happy," said Charity, "and I'm grateful."

"That's good," said Eugene. He felt better—steadier—now that he'd eaten.

"So," Jake said, "except for not liking the boat and the people on it, what do you think of the day so far? Glad you came?"

"I already thanked you. I already did that."

"Hey, that was a straight question. There's plenty of people to bite without biting your friends."

"Is that what I was doing, biting my friend? I'm glad I came, O.K.?"

Charity reached across the table and touched his chin with one of her long, slender fingers. "You didn't tell me Mr. Blasingame was interested in a fighter, in owning one you could work with. I think that's just wonderful—for him and for you. And for Pete, of course."

"The deal ain't done yet," said Eugene. "So far, it's only talk."

Jake said: "If Mr. B says the deal is done, it's done. You can put his word in the bank."

The four old black men had swung into a slow, waltzy kind of jazz number, and a few couples were swaying together across the floor to the music. Charity caught Eugene's hand in hers. "Dance with me."

"Not hardly," Eugene said.

Jake uncocked her hip, leaned and took Charity's hand in hers. "Come," she said. "I'll dance with you."

Charity hesitated for just a second, and Eugene thought he saw her cheeks darken with blood, before her lips parted, smiling, over her sharp, perfect teeth, and she rose to be embraced by Jake. They moved to the music, out over the floor, moved gracefully and naturally, and Eugene couldn't help thinking, as he watched Jake's narrow-hipped back, what a lovely couple they made. He felt a moment of uneasy fear for Jake. She no doubt thought she was the odd one out, when in fact she held, in the person of Charity, a bundle of snakes. But Eugene felt Jake could take care of herself, and he didn't want to think about it.

He watched the door through which Pete had disappeared and waited for him to return. But he did not. He started to have another drink, but thought better of it. His head had cleared and he remembered the urgency and anger in Pete's voice. The number the musicians were playing ended, and they segued into another. Jake and Charity remained on the dance floor.

A waiter came by the table and Eugene caught his arm. The waiter stopped, turned toward Eugene, balancing the tray he was carrying.

"The greenroom," said Eugene.

"You wish something, sir?" The waiter had an accent that he did not recognize.

"My friend went to the greenroom. Where is it?"

"I'm afraid I don't understand, sir."

Eugene stood up. He still had the sleeve of the man's jacket caught in his fist. "Take me to the greenroom, cocksucker." He had spoken louder than he meant to and realized, for the first time, that a tiny fear had been gnawing at him. Eugene had

188

not brought Pete here to drop him into something he could not handle.

"Please, sir. The cursing . . . Mr. Blasingame."

"Take me now to the greenroom or I'll tear your head off and puke down your lungs."

"Just let me get Mr. Reeker, he'll know . . ."

"Purvis already told me. And he's got my friend down there."

Eugene saw the waiter's eyes lift and focus on something over his left shoulder. When he turned to look, Eugene saw Mr. Blasingame across the room nodding his head to the waiter. He gave Eugene his tight, lipless smile and raised his carrot juice in toast.

"If you would follow me, sir," the waiter said.

"Why the fuck couldn't you save me all that to start with?" said Eugene, as they stepped out of the room into a long corridor.

"You should not curse here because . . ."

Eugene reached out and slapped the waiter hard on the back of his head. The waiter flinched and jerked his head into his shoulders but did not turn. "Shut up," hissed Eugene. "Don't talk to me." He didn't like any of this. Something smacked of kink here, and Blasingame had promised him that kink was not what this trip was about. Well, son, he thought, ain't the first time you been lied to, is it?

The corridor took them to the very back of the yacht, where they descended a narrow iron stairwell. At the bottom of the stairwell was a pale green door. The waiter stepped to the side of the door and averted his eyes.

"The greenroom, sir." The waiter turned and Eugene could hear his feet on the stairs as he rushed away.

Eugene opened the door and stepped inside. The room was not only painted pale green but it was lit indirectly from some secret source in an even paler shade of green that caused the heavy chairs, the couches, and the people moving through the

189

dim light to glow like phosphorus on the sea at night. There were no windows. Eugene stood very still just inside the door, waiting for his senses to adjust. Then he felt, rather than saw, someone at his elbow. When he turned, Purvis loomed there, enormous, glowing, with a flaking cream pie in either hand. Some white substance was smeared on his lips, and his powerful jaws worked and worked before he finally swallowed.

"Hello, Knockout," he said.

"Eugene, goddammit. Eugene Biggs." His throat was so tight that he did not recognize the voice as his own.

Purvis looked at the ceiling, and raised a pie. "Up there, you're Eugene. Down here, you're Knockout. And don't get superior with me just because I'm heavy. Heavy people need love, too."

"Jesus," said Eugene, "this is a bad dream."

"This is everybody's worst dream," Purvis said around a mouthful of cream pie, chewing, smiling. "Your worst dream come true. Mr. B is a dreamer, a purveyor of dreams."

He looked at Purvis, and the expression on Eugene's face made Purvis take a step backward. "You're not going to run amuck are you?"

"What?" Eugene said.

"Amuck. You are not going to run amuck." It was a statement now, not a question. "Because Mr. B does not like amuck. He will not deal with amuck. So don't. It's too far to swim home."

The phosphorescent figures moving through the green light had become more distinct now, recognizable. Across the room a tiny man dressed as a jockey was riding a wooden rocking horse. At regular intervals he savagely lashed the horse with a leather whip. A group of people stood in one far corner, and as Eugene watched, they parted briefly, and he saw Russell Muscle, naked, his body oiled, flexing in front of a huge mirror while the young man he had seen earlier at the end of the dock, tall and lean and very blond in his bleached-out and

190

wrinkled denims and tuxedo jacket, directed a long-snouted camera with one hand and waved wildly with the other.

"Yes, I know," said Purvis in a sympathetic voice. "Always something of a shock the first time. But you'll get used to it if you want to." He took another bite of the last remaining pie. "Or need to. But do try to remember that it is not Mr. B's pleasure to force anybody into anything. Or even to suggest, for that matter. He only provides the opportunity. Opportunity is his pleasure."

"Where is Pete?"

"You don't want to know."

"I came down here, didn't I? Don't hurt yourself, Purvis."

"I'll not, and neither will you. You're in our game now. This is our arena. And, yes, as a matter of fact, you did come down here of your own free will."

"Show me Pete, you fat tub of guts. I'm tired of talking."

Purvis said: "You've already established what you think of my body, Knocker. I do wish you didn't feel compelled to make continual reference to the fact that I'm heavy."

"You've gone way past heavy, Purvis. You've gone all the way to something else."

"There you go again," said Purvis, but he turned as he spoke and Eugene followed him past the tiny man lashing the rocking horse, to a section of the room partitioned off by a bamboo screen. Purvis stopped, licked his fingers, and pointed.

"Through there. And remember, amuck will not be tolerated."

"Eat shit and die," said Eugene.

Eugene parted the bamboo screen and stepped through. It was a small room and dimly lit by a small, shaded green light on a low table beside a narrow bed. Josephine sat on the edge of the bed, her face slack as though she might be asleep, even though her hooded eyes were open. Her left arm was tied off above the elbow by a looped belt, the end of which she held in her teeth.

Pete stood in front of her, holding a hypodermic syringe raised toward the dim light. He turned, glanced at Eugene, then back at the syringe, and thumped it with his finger. Eugene could not speak. He knew what he was seeing but looking dead at it could not believe it. He moved closer.

Pete bent and took Josephine's elbow and inserted the needle into the bulging vein there in the crook of her arm. He drew back the plunger and the syringe filled with blood. Eugene had seen more blood in the ring and in the gym than most men saw in a lifetime, but looking at the blood in the syringe, he felt his gorge rising.

"I'm in, darling," Pete whispered softly.

Josephine opened her mouth, releasing the belt, and as it loosened on her arm, Pete slowly depressed the plunger. Josephine's mouth opened wider and she licked her dry, cracked lips.

In a distant voice, she said, "Oh, God." She watched the syringe. "Boot it," she whispered.

Pete drew back the plunger, filling the syringe again with blood. He slowly depressed the plunger until it was empty.

"Again," she said.

He drew the plunger back and when the syringe filled with blood, he depressed it again.

"Once more," she said, her voice dreamy. "Just once more."

"You've got the hit," Pete said. "You're just playing now."

He withdrew the needle, placed a piece of cotton where the needle had entered the vein, and pushed her hand up against her shoulder to hold the cotton in place.

In a barely audible voice she said: "Thank you. Oh, thank you, God."

Pete placed one of his hands on the back of her head and eased her down on the bed, where she lay smiling, her eyes half closed.

"Ahhh, Pete," Eugene said.

Pete said: "She'll be all right by the time we get back to the marina."

"She ain't going to be all right for a long time, man . . . if she ever is."

"Well, at least now you know."

"I don't know shit," said Eugene.

"None of us do," Pete said.

Chapter Eleven

As Pete had said she would be, Josephine was all right by the time they got back to the marina. If anything, she looked and acted very tired, but they were all tired. The ride back into New Orleans was quiet, and everybody seemed happy. Pete and Eugene had been advanced two weeks' salary apiece, and they were to pick up a Buick Riviera from Mr. Blasingame's leasing agent the next morning, and had decided between them that the first thing to do was go see Bobby Barfield. They had not talked about what Eugene had witnessed in the greenroom. Eugene did not know what to say, where to start, and Pete did not seem inclined to say anything. The sight of the blood filling the syringe had sobered Eugene and scared him. The thought of anybody putting a needle in his arm was a nightmare to him, had been ever since he lived in the Hotel Rue and seen junkies with arms like bloody hamburger from shooting coke every fifteen or twenty minutes around the clock, that and watching them carried out on sheet-covered stretchers.

Charity drove and Eugene sat in the backseat with Josephine and tried not to imagine what her arms looked like under her long sleeves. Charity pulled up to the curb at the apartment over the shoe store. She was radiant from the sun and her hair shined silken.

"An absolutely gorgeous day, everybody." She looked from Pete to Josephine. "Don't you two guys be strangers."

"Hell, me and your man's partners. No way we can be strangers," Pete said.

"It was great fun today," said Josephine. Her speech was slurred. "I guess I drank a little too much."

"It was celebration," said Charity. "It's allowed."

Pete helped Josephine out of the car. "Eight o'clock in the morning, Eugene. Deal?"

"Deal," he said.

When they got back to the apartment on Exposition Boulevard, Eugene, for no particular reason, wanted a drink. It was not as though he was about to go perform, or that he had just performed, or that he was driven to the point of nervous collapse as he had been when he had got into the filing cabinet and read the terrible things Charity had written—no, he just wanted a drink. But the more he thought about it, the more he knew it was because he was worried about Josephine. If she was a stone junkie, what would that do to Pete? More than that, what would Oyster Boy do to Pete, finally make of him? Pete thought he knew more, understood more, than he really did. And at the same time, Eugene was concerned for Jake. He genuinely liked her. He thought despite her twisted life that she was good people. And now Charity knew her, had talked to her, danced with her. And Charity was trouble. Charity would break your bones and not even mean to do it, not even know she was doing it.

He sat on the edge of the bed and listened to Charity in the bathroom, the shower running and running. He had made the decision to cut himself loose from all that he had become since Budd abandoned him, simply say no to all that his life had become, but then he had introduced Pete and Jake into the midst of the same life. Pete to Mr. Blasingame, and Jake to Charity. He did not like it at all, felt the danger in it and the shame. But he told himself that it was not the way he had meant for things to work out, that he was only trying to save himself, and that both Jake and Pete had been around the block and were capable of fighting their way out of whatever they got into. To hell with it, he could not save the whole world.

"Why so deep in thought?"

Charity was standing in front of him, a towel wrapped around her. She smelled of body oil. He had not heard the shower stop.

"For some fucked-up reason, I was thinking about making myself a drink."

She watched him with a relaxed, easy smile touching her lips. "If you want to relax, I can help you."

"No," he said.

"A fuck will take the edge off the day," she said.

"I don't think so."

"Would you like me to make you a vodka tonic?"

"If you don't mind."

"Of course I don't mind."

She went into the kitchen and momentarily was back with a tall glass. He sipped it, and it burned his throat like raw vodka.

"How is it?" she asked.

"Strong."

"I thought if you needed a drink, you might as well have one that would help you. I have to get dressed and get out of here. Much research tonight. It's been a memorable day and I would not have missed it for anything, but all play and no work is not the way I can afford to run my life. So I really do have to go do some work."

"Go then," he said, raising the glass again.

"Oh, baby, couldn't you seem just the least bit disappointed?" She reached and touched his chin.

"I'm disappointed enough," Eugene said.

"That's better," she said. "That's my man. You'll be all right here tonight, though?"

"I'll be fine." The drink was helping and he felt a little better. Things would work themselves out. They always had.

"You'll sleep like a dead man, darling, after all that sun and salt water and good food. And you really ought to. You and Pete have a big day tomorrow."

He didn't answer but walked over to his desk, sat down, and picked up one of his daddy's letters, turning it in his hands.

"This deal with Mr. Blasingame," she said, "is a godsend."

"I'd just as soon wait to see how it works out."

"It'll work out fine. I know it."

"Most of the things I've known was gone be fine had a strange way of going sour on me."

She came over and kissed him on the side of the neck. "You get a good night's sleep. Things'll look better in the morning."

She dressed quickly and told him she would be back in a bit, but she did not come back that night. And he knew she wouldn't. He was glad for that, glad to be alone. For a while he sat at the desk reading over a stack of his daddy's letters, even tried to write one, but could not. He didn't want to write any more lies, would not write any more, and without the lies, he found he had nothing to say. He thought that with the money he had been sending home his folks were all right for now. But five hundred dollars a week was not going to cover the nut. If they got a good boy though, and won a couple of quick fights, he'd be making another hundred, and with two thousand a month, he could still send something home that would help. His daddy had not mentioned what kind of crop year he'd had, and that was not a good sign, but maybe he had other things on his mind—his brother's broken arm, his mother's illness—and simply forgot to mention it. He hoped so.

The solid sleep that Charity had said he would get did not come. But he managed to doze off just after midnight, after wrestling with the bedclothes, and getting up twice more for a stiff drink. When he did go to sleep, it was fitful, and he woke up tired and was barely dressed when Pete knocked on the door.

"I slid by and picked up the car," Pete said, exuberant, his hands and feet moving in a happy little dance. "Let's get in that sucker and get in the street."

They took the River Road from New Orleans to Baton Rouge instead of the interstate. It was not the shortest or the quickest way but it was the most enjoyable. They were in no hurry and neither of them felt like fighting the traffic.

Pete looked over at Eugene, who was driving. "Man, ain't we looking good? Ain't we pretty? Setting up here on these fine wheels, money on our hip, AC blowing. We ain't field niggers no more, captain. We don't take no shit from no swinging dick."

By now, Eugene was feeling good, too, his mind easy. "This thing drives like a new car."

"Brand new," said Pete. "Ain't nobody ever farted on these seat covers. Had to go to the mat with Blasingame's lease man, but look what I got for us to impress the citizens with."

"You missed your calling, Pete. You should have been a politician."

"I may be yet, son."

As they drove toward Baton Rouge, the Mississippi River was on the left, its banks lined with chemical plants, and on the right sugarcane fields stretched away to the horizon. Plantation house after plantation house, ornate, columned, invariably white, rose up before the fields of cane. Small cities of slave quarters stood behind the plantation houses.

Eugene watched the grayish clouds of fumes hanging above the chemical plants. He was glad for the enclosed car and the air-conditioning so that he didn't have to smell them. "I hear there are twenty-six of those things," he said, pointing to a chemical plant they were passing, "on the river between Baton Rouge and New Orleans. The fish must think they've died and gone to hell."

"Mississippi River ain't nothing but a ditch," Pete said. "You fell in that son of a bitch, they wouldn't even find your bones. That water would eat you up."

They passed a particularly enormous plantation house. A tiny row of slave houses stood right beside it. Pete turned to stare at it. "Them right there is where the house niggers stayed, don't you think?"

"I don't like to think about it."

"I heard," said Pete, "that one of these suckers somewhere in here owned six thousand slaves. What you think it'd feel like to own six thousand people?"

"Pretty shitty, I hope."

"Naw, man. You got that wrong. Owning six thousand people would feel good, gooood."

Eugene watched the road, where the heat waves rising made the light a shimmering of the air.

"How many people you think Blasingame owns?" Pete said.

"I don't think about it."

"You best do. I bet six thousand wouldn't even start the count. And yesterday he just raised the number by two."

"You're a good guy, Pete, but you full of shit sometimes. He don't own me and he won't ever."

Pete smiled. "You just ain't thinking. You got his money in you pocket, don't you? You driving his car, ain't you? That's his AC cooling you balls. Oh, yeaaah, he got him two more yesterday."

"You ain't been where I been or you wouldn't talk shit like that," Eugene said.

"Your memory ain't good, man. You done took yourself out too many times. Forget what I told you yesterday on that boat? I been owned, lock, stock, and balls. And let me tell you, the guys that owned me felt so good, it was hard for them to quit dancing. Yeah, dancing and smiling. So don't tell me about owning and being owned. Shit, I been owned all my life." His expression had changed while he talked. He first looked baffled, then angry. "And from what I seen of your life you been on a leash, too."

Eugene reached over and put his hand on Pete's shoulder, a touch gentle as a lover's. "I dig, man. But leave it alone. We're together. We're going to do it. Let it slide."

Pete looked away toward the cane fields. "I got blood on my mind this morning, Eugene. World don't ever seem to git right, does it? Git right so a man can move?"

"Yeah," said Eugene looking straight ahead.

They drove in silence for a while.

"Want the radio?" Pete said.

"No."

After another long silence, Eugene said: "Tell me."

Pete looked at him and then away toward the cane fields so Eugene could not see his face. "About Tulip?"

"Yeah."

"Freaked you more than some yesterday, didn't it?"

"Enough."

"I probably should have told you before, but I had her clean, man, I had her clean." He looked back at Eugene. "You don't have to listen to this if you don't want to hear it."

"Talk. I don't mind." But he did. The last thing in the world he needed was to listen to this. But Pete was his friend. And he thought that Pete needed to talk.

"What do you think a pretty girl like her, young as she is, is doing fucking a teddy bear in front of the safe, churchgoing citizens from Atlanta and St. Louis? The spike, man. She's in love with the spike."

"Yeah," Eugene said still looking straight ahead.

"And I'm in love with her."

"What you going to do?"

"Get her clean again, man, and keep her clean. That thing on the boat, that was just a standing eight count. She'll come back. She's a champ."

"I hope you right, Pete, I do. I like her. She seems like good people, and I ain't met many since I been in New Orleans."

Pete stared out the window in silence and then in a quiet voice, almost as though talking to himself: "How the hell was I supposed to know they might be dope and works to do it with on a boat owned by a man like Blasingame?"

"I should have told you."

Pete's head snapped from the window to look at him. "You knew?"

"No."

"Then what the hell you talking about?"

"I should have told you it was subject to be anything on it, anything where Blasingame is concerned."

"You know something I don't know?"

"Yes."

"What?"

"I told you he was a shark."

"That don't tell me nothing."

"You don't want to know. It wouldn't do you any good if you did."

"Now ain't this some shit," Pete said.

"You said you'd been in the water with sharks. Anytime you get close to him and leave, put one hand on your ass and the other on your head and make sure they both still there. Remember I told you so."

"Tell me the story on the greenroom. What was that all about?"

"I can't. Didn't know it was there. The first time I saw it was the first time you saw it. But I'll tell you this. It didn't surprise me when I walked into it."

"That was some kind of sick."

"Yeah."

"Did you see the guy who thought he was a jockey? Whipping up on that rocking horse?"

"He might have been a jockey. He might be a jockey. And, yeah, I saw it."

"And the naked iron freak."

"I saw all of it, Pete. I walked through it to get to you, remember?"

Pete reached over and put his hand on Eugene's shoulder. It was the same gesture Eugene had made earlier to Pete. "It's one thing about it, boss man. You and me don't have to whip no rocking horse and we don't have to get naked for a camera. Fuck'm and feed'm fish. We on the way to find us a fighter. You cover my back and I'll cover yours. They can twist their own heads, but they ain't twisting ours."

"Maybe."

"What you mean, maybe?"

"Power is power. And Blasingame's got it. A man can get twisted quicker than you think. Power is nothing to fuck around with."

"I can dig it. Like I told you, I don't care about Blasingame or which tree he swings from. I could give a shit. He got me out of the Flesh and Flash. That's all he means to me."

"How bout we drop this," Eugene said. "It's giving me a headache. Let's go talk to Bad and see if we can find us a fighter. Deal?"

"Deal."

Pete reached over and turned on the radio and they rode the rest of the ninety miles to Baton Rouge in silence. Eugene did not know the city, and Pete had to guide him through the twisting streets and heavy traffic to the gym, which was upstairs over a warehouse. A small sign no more than two feet square hung over the door leading to the stairs: BIG JOE'S FIGHT CLUB. Big Joe Rocker had fought as a heavyweight, but he had no speed and, surprisingly, because he was notorious for his strength, no punch. So after twenty fights, with a ten and ten record, he put on a hundred and fifty pounds of hard fat and muscle and started wrestling. But his heart was in boxing, and when he retired from wrestling, he opened a fight gymnasium. He found Bobby Barfield on the street—a mean, tough, fearless kid with two felony arrests and a string of misdemeanors on his police sheet already—and put him in the ring, talked to him, fed him, loved him, and made him the fighter he was today. But Bobby Barfield was still the kid he had found on the street: mean, tough, and fearless. The only thing that had changed was there had been no more entries on his police sheet.

Going up the stairs they could hear the staccato rhythm of speed bags and the heavy, irregular thumps of gloves on the heavy bags, and smell on the heavy air a mixture of sweat and liniment and rosin. The gym was enormous, poorly ventilated, and already hot as a steam bath at ten-thirty in the morning. In the center of the big room was a ring, and in the ring, Bobby Barfield was sparring with another fighter who looked to outweigh him by twenty-five pounds. The sparring partner wore a headgear; Bobby Barfield did not. Barfield was wearing gray sweatpants with the cup and leather belt on the outside of them and a black T-shirt on the back of which was printed in block letters: BAD AND BEAUTIFUL, and underneath that: LET'S GET IT ON. Joe Rocker leaned on the ring apron with a stopwatch in his hand.

"That's Radlovich Bad's going with," said Pete. "They must have brought him in from New York."

Thomas Radlovich was only a few fights away from being ranked as a heavyweight and was himself training for an important match at the Garden in New York. As a light heavy

Barfield was quicker and, everybody agreed, the heaviest hitter in his division. Plus, he had heart. He had never been off his feet and he had never been stopped. As he said himself, not often and never in a voice that could be interpreted as that of a braggart, "You'd have to kill me to stop me."

In the ring now, Radlovich was muscling him pretty good, bulling him around, leaning on him and working strong to the body, as he had no doubt been instructed to do. Eugene and Pete walked over and leaned on the apron beside Joe Rocker, who glanced at his watch and shouted instructions. He didn't even notice them.

Radlovich had Barfield caught on the ropes, pinned, and was holding him there with the power of his punches. Eugene winced every time Radlovich worked to the ribs in short, brutal hooks, his head down on Barfield's chest.

"Turn'm, Bobby!" shouted Joe Rocker. "Goddammit, spin out of there. Get off those fucking ropes!"

Every time Radlovich hooked, he grunted, and he kept on hooking and grunting, using his weight. Barfield caught a lot of the punches on his elbows but he could get no leverage. Both men were washed down in sweat.

"Time!" called Joe, but Radlovich kept hooking. "Time, goddammit, time!"

Radlovich stepped back and glanced over to where they stood, his eyes distant, as though waking from a trance. Eugene knew that he had been so locked into what he was doing, into combat, that it had been a kind of trance, a fierce, dreamlike state into which not even pain could enter.

When Radlovich turned and walked back to his corner, Barfield stayed where he was, leaning against the ropes, his face a mask showing nothing, except for the eyes. His eyes were unmistakably alive with something that clearly said: kill, destroy. He slowly walked to his own corner where an old black man, very tiny and toothless, with a cap of white, tightly curled hair, took out Barfield's mouthpiece, gave him the water bottle to let him rinse his mouth out, and toweled him off. The old man, whose age nobody knew, not even himself, was named Tater Jordan. He talked constantly while he toweled Barfield, but Barfield didn't listen, and Eugene knew why. Tater was talking to himself, probably giving instructions to himself in one of the fights he himself had fought long ago. He was punch-drunk now, his brains scattered, his speech slurred. But he was the best cut man in the business. He could stop anything that bled. Bobby Barfield once told Eugene that you could get run over by a Greyhound bus and Tater could stop every bleeding spot on you in one minute, the time a fighter had between rounds. He's a one-minute man, Barfield had said, and he ain't nothing but magic, a magic one-minute man.

In the corner, Barfield did not sit on the stool, but stayed in a little bouncing dance on his toes. Across the ring from him, Radlovich, sitting on the stool, breathed heavily. Barfield glanced down at Pete and Eugene, his fierce eyes burning, unblinking, and did not even seem to recognize them, but turned to stare across the ring at Radlovich.

"How is he?" said Pete.

Joe Rocker looked at them for the first time. While the fighters had been going at it in the ring, veins had stood in his forehead and in his neck, and a little slobber had dripped from the corners of his mouth, but now his face was relaxed, smiling, and he was as calm as an old lady crocheting in a rocking chair. "What is it, kid?" he said to Pete. "Good to see you, Eugene. It's been a while. Too long. What brings you to the gym?"

"This and that," said Pete, and then again: "How is he?"

"The Bad Man is settin dead on ready. He is naturally ready to torment and terrorize. He's ashes and sackcloth, Satan, sin and death. I would sho hate to be settin over there where Radlovich is settin." Then he looked up into the ring and shouted, "Time!"

When the two fighters came out, Radlovich tried to close with Barfield, use his weight muscling him again, but he couldn't catch him. Barfield stayed outside, moving left, then right, in and out, and circling to his right, his left hand straight and working like a piston, powerful enough to snap Radlovich's head back every time it landed. For nearly thirty seconds Barfield stuck and moved, throwing not a single right hand, playing with the heavier man, and then he stepped inside and double-hooked with his right hand to the heart and unleashed a left cross that, although it caught Radlovich high on the side of the head, dropped him flat on his back, where for an instant Radlovich did not move, not even a twitch. Then he slowly sat up, shaking his head, trying to clear it as he went to one knee and started to rise. He was a tough kid, thought Eugene, and he would have beaten a nine count. But he had had more than enough for today, and Joe Rocker knew it.

"Time!" Joe called. Radlovich was on his feet now. Tater was already through the ropes, taking his mouthpiece out, his headgear off. "Good work, Thomas," Joe said. "Really good work."

Bobby Barfield softly tapped Radlovich on the chest with his gloved hand and then dropped his arm across his shoulders. "Great workout, man. You damn near broke my fucking ribs."

Radlovich smiled. He had blood on his teeth. He said something but his voice was too shaky to make it understandable and when Barfield walked him to the corner, his arm still across his shoulders, Radlovich's legs were unsteady. After Radlovich was on his stool and Tater was ministering to him, breaking an ammonia capsule under his nose, Barfield strolled across the ring and leaned on the ropes, looking down at Pete and Eugene. He did not look tired at all, and neither was he breathing heavily. Except for the sweat running down his body, he might have just come in from a walk in the park.

"Hey, bro," he said. "Come down to see what the real thing look like?"

"You looking at the real thing, sucker," Pete said.

Barfield looked around as though people were standing on either side of him. "Listen to this! Listen to this fool talk."

"You looked great, Bad," Eugene said. "You looked really great."

"The Bad Man be bad," he said.

Pete looked at Joe Rocker. "Caint you git this jiveass nigger to wear a headgear? Damn!"

Joe only shrugged and smiled.

"Never be cut in my life," said Barfield. "Not even shaving, little man."

Eugene knew that a headgear did nothing to protect the brain, nothing to keep a fighter in training from getting knocked out, but it did keep him from being cut from taped laces or an accidental head butt. He thought it was stupid, in any case, for Barfield to be sparring this close to a championship match without a headgear. But that was the Bad Man. He was notorious for doing things his way.

"Listen," said Barfield, "when I meet the man can whup my ass, he be able to tell me errything to do. But that muthfucker, he ain't borned yet. Listen, I gotta git a shower. You be around for a while?"

"For a while," said Pete.

"Later," said Barfield, stepping through the ropes and dropping lightly to the floor.

Joe said: "Less us go drink us a Co-Cola."

They followed him across the gym and through a little hallway to a tiny office. When he opened the door, a blast of refrigerated air met them. He went to a scarred desk covered with scattered papers about a foot deep that looked as though they hadn't been touched in a month. Pete and Eugene sat on a leather couch that had jagged tears in it from which spilled black padding. There was no carpet on the wooden floor, and yellowed fight posters, some of them going back twenty years, were thumbtacked from ceiling to floor like wallpaper.

Joe reached behind him and brought three Cokes out of an ice chest. He opened them and Eugene got up and took one for himself and one for Pete. Joe took three pills out of a bottle and swallowed them with Coke.

"Darvon," he said. "I keep a headache. I just hope my stomach don't go before we get through this goddam fight."

"Ah, you got it in the bag," said Pete.

"They ain't nothing in the bag when you going against Cobra Carnes. But less us talk on something else. I don't need to think about it. What you here for? I know you ain't drove ninety miles to drink a Coke with a fat wrassler."

Eugene started to say something and then decided against it. Pete was better at talking than he was.

"You got that right," said Pete. "We come to talk to the best fight man in the business."

Joe Rocker's heavy face lifted in a tired smile. "You just trying to help my headache, but I prechate it, I do."

"We looking to get back in the game," Pete said.

"We thought you could point us toward a boy that might have something," Eugene said. "Somebody we could bring along the right way, maybe all the way. But we ain't playing. He's got to be the goods."

Joe Rocker was incredulous. "You want to own a fighter?"

"Now you got everything right," Pete said.

Despite the air-conditioning, Joe Rocker was sweating. He wiped his forehead with a soiled handkerchief. "Jeez, the last I heard, Pete, you was . . ."

"I ain't there no more. We dead serious, and we got the money. We got all the money and all the time we need."

"Did you fucking guys rob a bank?"

"We got lucky on a number," Eugene said.

Pete looked at him and smiled. "Yeah, a number."

"You could be out a hundred thou before you ever see a penny."

"Come on, Joe, I was seven in the world, goddammit, I think

I know what it costs." Pete's face had gone tight and his voice was hot. But then he relaxed and his smile deepened. "It was a very big number. Can you help us or not? I mean we ain't looking to do this today. Give us a name or names. We can go to Philly or New York or wherever and do what we have to do."

"You got that kind of money?"

"I told you the number we hit was very big."

Joe reached behind him and got another Coke out of the icebox. He opened a drawer of his desk and brought out a bottle of black label Jack Daniel's. "You boys want a drink?"

"We'll pass," said Eugene. "I thought you had a bad stomach."

"What hurts my stomach helps my head."

"I heard that," Pete said. "Same with me."

Joe drank off several enormous swallows, his sweating throat working under the upturned bottle. The color rose in his face and he leaned back in his chair and closed his eyes. He sat very still for a very long time, saying nothing.

"You gone to sleep, man?" Pete said.

"I'm thinking."

"What you thinking?" Pete said.

"I'm thinking I know the boy you want."

"But you got him locked up, right?"

"Right."

"Well, like I said, we ain't looking to do this today."

"Boy's name is Jacques Deverouge. Cajun out of Lafayette. Unschooled. Awkward. And a potential champion."

"I magine," said Eugene.

"He'll be in this afternoon when he gets off work."

"What's he work at?"

"Janitor."

"How old?"

"Nineteen."

"Ever beat anybody?"

"Three times Louisiana Golden Gloves champion."

"Wonderful," said Pete.

"I don't think we interested," said Eugene.

"You ain't seen him. And besides, you caint afford him."

"How many fights he had for you?"

"Three."

"Win any?"

"All by knockout."

"What division?"

"Light heavy. But he'll grow into a heavyweight."

"Shit," said Pete.

"Damn," said Eugene.

"I ain't been thinking to sell his contract."

"Then why we talking?"

"You said the number you hit was big."

"Who'd he fight?"

"Three no-talent nobodies."

"Figures," said Eugene.

"I wouldn't think about selling him, but I got no time to do the kid justice," Joe Rocker said. "I keep a headache and my stomach is killing me." He raised the Jack Daniel's bottle again and bubbled it. "It'd cost you fifteen thousand to git him, if you want to know."

"I didn't ask," said Pete. Then to Eugene: "Did you ask?"

"Not me."

Joe Rocker smiled. "Wait'll you see him."

"He caint be shit," said Pete, "or you wouldn't have him on the block."

It occurred to Eugene that they were talking about this kid as though he was a horse or a valuable dog. A terrible game, he thought, that I'm lucky to be out of.

The door opened and Bobby Barfield came in. He was resplendent in powder blue: trousers, open-necked shirt, and jacket, all silk, with three heavy gold chains around his neck.

He stood with his huge, heavily muscled hands palm out and turned in a slow circle.

"Ain't I pretty?" he said.

"If you was a woman," said Pete, "I'd put you on the street."

"Come on, I'll buy you dicks some grease. I gotta eat, grab a nap, and be back in here beatin the shit outta somebody at five." He looked over at Joe and said: "Put that fuckin bottle back in the drawer, Joe. For Christ's sake, you gone die, do it on your own time. We got the world to win. You think Cobra's manager be consortin with Mr. Jack Black? Damn!"

"My head hurts, Bad."

"Hurt? Hurt you say? Sheeit! What it be I come back to the corner and say, boss man, I hurt? What it be? Lose that bottle!"

Joe put the bottle back in the drawer. Bad came around the desk and put his hand on Joe's thick shoulder: "I love you, and we gone win the world together. Come on, we buy these two no-counts something to eat."

"I got things to do here. Besides, my stomach don't feel good. How was your weight? You weigh?"

"Notch-solid. Errything on, above, below, and around Bad and Beautiful be notch-solid. Can you dig it?"

"Git the fuck out of here," said Joe Rocker.

Chapter Twelve

Eating lunch with Bobby Barfield was like eating in the middle of a parade. They had arrived at Toutonte's Cajun Place in Barfield's solid-gold Cadillac with Barfield himself driving as though intent on committing suicide before sundown, in and out of traffic, running stop signs, making the tires scream at every corner, all the while keeping up a running chatter without apparently watching the road at all.

"I wish I had a taste of whatever you been eatin, Bad. Dammit! Son, slow this thing down."

"You ever see the Bad Man do anything slow? Hell no! Errything I do, done full bore. Besides, these wheels famous as God in Baton Rouge. Nobody gonna fuck with us."

"Except maybe death," Pete said.

"Death show up, I kick his ass, that's the long and the tall of it."

Toutonte himself led them to a table in the center of the huge restaurant. Framed fight posters hung across the mahogany back bar and lined the walls, many of them of Bobby Barfield, some posed in the gym, some of them action shots of him destroying some other fighter. A whole covey of young waiters followed them to the table, all of them black and out of themselves with excitement now that the Man was here, shucking and jiving ninety miles to the minute.

"The dancing man!"

"Lookit that neck flash!"

"Bad be holding some gold!"

"Bad Man, how many rouns? Say how many rouns!"

"Yeah, let us git something down on the roun."

"Buy us some chain-and-street-corner flash."

Bobby Barfield only smiled and waved his thick hands and nodded his head acknowledging the homage he was being paid, his due, the celebration that was his and only his. When they sat down at the table, Toutonte, who had been indulgent of the waiters, beaming himself, looked up once, his smile finally gone, and the waiters were gone as rapidly as they had appeared.

"The usual, Bobby?" said Toutonte.

"The usual," he said, and offered his hand. He and Toutonte did a power shake, and Toutonte looked at Pete and Eugene. "Would you gentlemen like a menu?"

Pete said: "Whatever is good enough for Bobby is good enough for me. The same."

"I know that's right," said Barfield, signing an autograph for two young, very pretty white women who had appeared at the table.

"Do it three times," said Eugene.

"Very good," said Toutonte.

Bad watched Toutonte walk away, moving—seeming to glide rather—on the balls of his feet, even though he must have been fifty or more.

"Hell of a welter in his day," said Bobby. "Never saw him go myself. But he ain't nothing but a fucking legend. Never won the world, though. Come close, but close don't count in nothing but hand grenades." He looked back across the table. "How did it look for Cobra when I was going in the gym today?"

"I don't know. How lucky are you?" said Pete.

"Lucky? Sheeeit! Bobby Barfield ain't lucky. Ain't never been lucky. Luck is for losers. Now, you was lucky. You was a very, very lucky dick."

"Carnes got no fucking chance," said Eugene.

Barfield studied Eugene across the table, his eyes hooded, no longer smiling. "You ain't holding bad shit for a white boy, Eugene."

"My shit was very good," Eugene said, "until it went bad."

Barfield glanced out across the restaurant. "Now, I heard that. The best shit in the world will go bad sooner or later."

"Mine went bad early."

"Happens," said Barfield.

Two fat men in three-piece suits came by the table. One of them touched Barfield's shoulder. "We in your corner, champ. It's all the way this time." Barfield glanced up at them, but didn't answer. When they walked away, he said, "Now what would them suckers know about nothing?"

"Nothing," said Eugene. "They wouldn't know nothing."

"Only thing wrong with you is you skin," Barfield said to Eugene, smiling now. "You all right, man."

Their lunch came, a steak so rare it rested in a little puddle of blood, a small, plain baked potato, a wedge of iceberg lettuce, steamed carrots and broccoli, and a tall glass of orange juice. The waiter, a young black man with a slight stutter and very shy, asked Barfield if he would sign a napkin. Barfield signed it and handed it to him.

"I'm beholdin," said the waiter, staring at the napkin.

"Ain't nothing but a name," said Bobby. "Take care of yourself."

The waiter walked away still staring at the napkin.

"Damn, Bobby!" Pete said. "Orange juice here in the middle of the day?"

"Bad don't bruise," said Barfield. "I work strong to the orange all day long. You done that, you wouldn't be showing tits and clits for tourists to jack off to now."

"I'm not," Pete said. "I'm outta there."

Forking a hunk of bloody beef into his mouth, Barfield said, "Say what?"

"Run it down for him, Eugene," said Pete. "You made it happen."

Eugene ran it down for him, told him pretty much what they had told Big Joe earlier. He talked slowly, trying to get everything right. Bobby Barfield did not interrupt, only glancing up now and then from his steak, chewing slowly, acknowledging with a nod or a handshake the people who came by his table. The waiter filled Barfield's glass three times with orange juice before Eugene was finally finished. Barfield was silent as he chewed and swallowed his last bit of steak. Then he pushed back from the table and stretched his legs.

"Say you got the money to back a boy," he said, "and I could give a shit where you got it—but that number jive won't fly, they ain't no number that big, or it was and you won it, I would have heard—but say you got the cheese to do it with, you know what looking for a champion, or even a contender, is like? You two gone feebleminded on me? But I guess you did, so I'm gone tell you for nothing. Looking for a champion is like looking for a bitch that gone love you for youself and you alone forever. Can you dig it? You got a better chance a goin to the fucking moon as one of them assnauts. Did you hear that?"

"We heard and we know," Pete said.

"Then we ain't got nothing to talk about," Barfield said.

"You know a kid named Jacques Deverouge?"

"A coonass," said Barfield. "Dumb as a rock!"

"Is he any good?"

"He's a kid, man. Nobody can know that. Or does know that. Or will know that for longer than you even want to think about."

"You ain't being helpful," said Pete.

"I ain't paid to be helpful, paid to fight."

"I thought you was my friend."

"Pete, you know you ain't got no friends. But, K.O., he's a big, strong kid with a good left hand, or what might be, he ever went to school. A right hand I've seen drop some pretty good guys in the gym. He's still growing. I think he got heart. Seem like to me have a lot of heart. Saw him go three rounds in the gym with his nose broke bad and Big Joe trying to git'm outta the ring. He's a willful boy, and blood don't scare him."

Eugene said, "That sounds like the guy we want to think about."

"Yeah, I could think some on that," Pete said.

"You think on it," said Barfield. "I'm thinking on a nap. The Bad Man needs some Z's. Catch Deverouge in the gym, middle of the afternoon. I'll drop you by on my way home."

When Bobby Barfield dropped them at the gym and then left thirty yards of rubber pulling away from the curb, Eugene said: "Driving the way he does, you much think he'll live long enough to fight Carnes?"

"He'll live. Don't think twice about it. The Bad Man is too mean to die."

"Had an uncle tell me that once. Got kicked in the head by a mule and died the next day."

"Happens," said Pete. "Come on. We'll try to find out how long it's gonna be before this kid shows."

As it turned out, they didn't have to wait long. They spent the better part of an hour in Big Joe's little refrigerator of an office watching him drink Jack Daniel's and Coca-Cola and

216

listening to him complain about his stomach, but then he looked up at an old, big-faced clock on the wall and said, "He's coming in right now. Kid lives to get in the gym. You can set you watch by him."

As they were coming out of the little hallway leading into the gym, Jacques Deverouge was coming in the front door carrying a canvas bag with his name stenciled in red on the side of it. The white T-shirt he was wearing was emblazoned across the front with THE BOAR COON. He was at least six four, lean and big-boned, with dark skin, intensely black eyes, and straight black hair that hung to his shoulders. To Eugene, he looked more Indian than he did Cajun.

"You right about one thing," said Pete, "he'll grow into a heavyweight all right."

Big Joe said: "He's coming in at a full light heavy now. He gets fed right and trained right, he'll muscle up to two thirty solid. I just ain't got time to do the kid right is all, else I wouldn't even talk to you about him."

"You got a big heart," said Eugene.

"What everybody tells me," Joe said. "You want I should innerduce you to him now or what?"

"We'll watch for a while. Just see how things go," said Pete.

Things went just fine to suit both Pete and Eugene. When the kid came out of the dressing room, he spent fifteen minutes in a blur of motion with a skip rope. Quick, good feet. Moved light. On the speed bag, he tripled with both hands and on the heavy bag his punches landed straight and hard and his hooks were good.

"He don't look bad," Eugene said.

"The bag don't hit back. Wait'll we see him go a few rounds in the ring."

Jacques Deverouge was supposed to go three rounds with a heavyweight, an over-the-hill ham-and-egger who was on the downhill slide of a great career but who could still hit. But he could not catch the kid. Deverouge was in and out, side to side, turning the other fighter at will, but he still got caught with a couple of very good shots. They didn't even make him blink. What they did do was make his eyes become more intense and a snarl appear on his lips even though his teeth were clamped tightly on the mouthpiece.

"He's mad," said Pete. "He's lost his cool. Ain't good."

"He'll get over that," Eugene said.

"Maybe."

Big Joe, at the ring apron, was about to call time when Deverouge stepped inside and threw a left hook that didn't travel more than eight inches, and the heavyweight dropped flat on his back and did not move, except for the erratic twitching of his left foot. Deverouge stood looking at the man he had knocked down, his face showing nothing, neither anger nor elation; then he walked away.

"What now?" asked Eugene.

"I'm going a few rounds with him."

"For Christ's sake, Pete, how long since you had a pair of gloves on?"

"I was seven in the world, sucker." Pete sounded angry, and Eugene knew he was.

"Was seven in the world. You probably two hundred and fifty-seven now."

"You got a bad mouth sometimes, Eugene. Damn lucky we friends."

"I ain't puttin the mouth on you. Just saying a fact."

"You ain't saying shit!"

Pete walked over to the ring where Joe stood holding the stopwatch. "Think your boy'd wanta go a few with me? He ain't hardly worked up a sweat yet."

"I know he would. And I'd be grateful if you'd work with him."

"I ain't doin it for you," Pete said as he walked to the dressing room.

Joe found some shoes, a cup, shorts, and a headgear for Pete and while he dressed out, Eugene watched him with a growing uneasiness about what might happen. He didn't want to piss Pete off, but he was his friend and he didn't want to see him hurt. That kid out there in the ring could hit.

"Pete, I'm sorry for what I said about being two hundred and whatever, but this ain't the way to go about it. I don't want to see you fucked up." Then: "I love you, man."

Pete stood up from lacing his shoes. "You ain't got to tell me that. I know. Sorry I got hot."

"But I don't understand. What's the point of all this?"

"If he can catch me and hurt me, we gonna buy us a fighter."

"We taking the first boy we look at? That don't make sense."

"The fight game don't make sense. But I can see and I know what I see. I know what I know. I need bread bad, man. I ain't got no time to fuck around. Now let's go out and see what the kid's got."

When they went out into the gym, Jacques was standing down by the ring with Joe. He left Joe and came to meet Pete.

"You really Pete Turner, you?"

"If I ain't, everybody's been lying to me. You wanta work a couple, three rounds?"

"You wan beeg hero when I'm going to be thees high, you," said Jacques, holding his gloved fist to show the size of a child.

"What's you best hand, kid?"

Jacques held up his right fist. "Thees the beeg wan, you. But"—and he held up his left hand—"thees wan, nobody know, you."

"Let's get in the ring and do it," said Pete.

They climbed into the ring, and Eugene stood on the apron in Pete's corner waiting for Joe to call time.

"I'm gonna try to set him down. If I get him with my best shot, and there ain't no way in hell he can keep me from doing it, and he don't go down, then he's holding a fucking chin."

"This is crazy."

"He can't hurt me cause he can't catch me."

Joe called time and for the first thirty seconds, Pete circled left and then circled right, with Jacques following patiently, moving just on the balls of his feet, nearly flat-footed. Pete was outweighed and he was giving up ten years to the kid, but he looked good. Jacques tried nothing fancy, was in fact showing great patience, calm, and confidence, and he did have a good left hand. It was straight from the shoulder. And it had snap.

Somewhere just after thirty seconds, Pete faked left, went right and caught the kid with two crushing right hooks, doubling to the head, and came back with a left cross that landed right on the point of the chin. Jacques should have dropped, but he didn't. He didn't even look stunned, only confused. Eugene could see it clearly in his eyes: Hey, this is supposed to be sparring. This guy is trying to take me out. What's the story here?

Pete was snapping the kid's head with left hands that seemed to be over before they started they were so fast and clean, and always he circled. But then he stopped, and Eugene saw him set himself as he moved in, heels flat on the canvas, and brought a right hand that had all of his body behind it. When it landed, it didn't move the kid at all, who came back with a counterleft that made Eugene know it was all over. Pete's knees buckled, a veil dropped over his eyes, and then he dropped, not slowly, but straight forward onto the canvas, facedown.

Pete didn't know who he was or where he was until finally he came around in the dressing room, lying on a training table, an ice bag on the back of his neck and an ammonia capsule under his nose.

Eugene was bending over him, and Pete looked up into his eyes.

"What did he hit me with?"

"A left hook over the heart. You all right? You must feel like shit."

Pete only smiled. "Feel like shit? Man, I feel great, super, wonderfuckingful. Damn!" He sat up on the table, and saw Jacques standing there at the end of it with Big Joe. Then he quit smiling. "Lucky shot," he said to Big Joe.

"Sure," said Joe. "I'll be in my office."

Jacques said, not without pride: "You taken and got a beeg wan, you." Pete only watched him. "I'm gonna take myself and put me in the shower now, you."

"Right," said Eugene and watched the kid walk away.

"Over the fucking heart." Pete shook his head in wonder. "Put me down with a shot to the heart!"

"The punch was luck, nothing but luck," Eugene said.

"The power wasn't luck, my man. This is the kid we going with."

"It don't make sense to go with the first boy we look at."

"How many times I got to tell you? Try to listen. The fight game don't make sense." Then he looked away toward something farther away than the room they were in, or the gym, or maybe even the city. "Nothing makes sense," Pete said quietly.

Chapter Thirteen

And so Jacques Deverouge came from Baton Rouge to live in New Orleans in Pete's old apartment above the shoe store on Prytania Street and do road work with Eugene in Audubon Park. Pete and Tulip moved into the Jean Baptiste Lemoyne Arms one floor below Jake's apartment.

"Into that fucking building on what you making?" Eugene said. "That ain't possible," said Eugene.

"Long-term financing," Pete said.

"You mean like from here to the moon, right?"

"Yeah, from here to whatever."

"That's long enough, but it still ain't possible."

But it was possible, as it turned out, because Mr. Blasingame owned the building where Jake lived and came to some arrangement with Pete which Eugene did not understand and refused to inquire about because he knew entirely too many things about too many people already, things that he had come to know—at least, in some instances—without ever quite knowing how he had come to know them. At times he felt himself tainted, rotten maybe, at the center of his life by the knowledge he carried about with him.

"Hey, what you had been thinking, you?"

Eugene glanced over at Jacques running beside him in the steaming early-morning grass of Audubon Park. He shook his head. "Nothing."

"You face all closed like un feest." He closed his huge left hand into a fist and pumped it in front of him before he double-hooked with it and followed with a right cross.

Eugene smiled. "Feel good, kid. Feel fine."

The huge, square white teeth flashed in Jacques Dever-ouge's dark face and he picked up the pace they were running. *"Laissez le bon temp rouler!"*

"Damn right, Jacques!" shouted Eugene. "Let the good times roll!" Eugene threw back his head and laughed with the sheer good feeling of being out in the bright, clean air of dawn, running over the damp grass with Jacques who was, as he always was, inevitably filled with hope and promise, strong and sure of himself. Anything seemed possible when he was with the kid.

Eugene's wind was still good as they turned into their fifth and last mile. It didn't hurt a thing that he was running in only shorts, T-shirt, and Adidas shoes while Jacques wore heavy combat boots and had a towel around his neck stuffed into two sweatshirts that were in turn stuffed into thick sweatpants. The pants were sweated through and dark wet stains were fanning out under his arms and between his shoulder blades. But Jacques was still running easily, his knees high, breathing through his nose, his mouth clamped shut.

If he concentrated on the kid, Eugene's life sloughed off his back like the skin off a shedding snake. He felt renewed, revitalized. He was able to remind himself that he had turned his life around, that for the first time since the dreadful fight in Madison Square Garden, he was on his way straight up instead of turning the same old confused circles. Everything was working better than he could have hoped for.

They had got the kid's contract from Big Joe not for fifteen thousand, but for forty-five hundred, which Pete had been quick to point out to Mr. Blasingame, who was equally quick to point out to Pete that he did not need it pointed out.

"I know what you got him for, Pete."

They were in Mr. Blasingame's office, having just driven directly back from Baton Rouge. The New Orleans skyline was lighted and blinking in the darkened ceiling-to-floor window there behind where Mr. Blasingame sat, his elbows on the desk, his pale, bony fingers tented in front of his face.

"I didn't know it myself," Pete said, "until two hours ago."

"Precisely the point," Mr. Blasingame said. "Two hours." He untented his hands and leaned back in his chair. "If we are to do business, Pete, you must understand that what I do not know, my people know. I pay them to know. Nothing I have an interest in goes unnoticed and unreported. Nothing." The gray, tired flesh of his thin lips pulled into a grin that made Eugene remember leashes and collars and fat oysters thick as a man's wrist. "It is my pride, Pete, that I know you better than you know yourself."

"I don't know what the fu . . . what you're talking about."

"I know that, too, Pete. But enough of this. I've had a long day. You've done well. Moved more rapidly than I thought you would have, or perhaps you should have. But I love nothing if not things done with great dispatch. I know the boy. I approve of your actions. It's a place to start. I bid you good evening, gentlemen."

As they rose to go, Mr. Blasingame said: "I shall be wanting to meet this young man."

"Hell, I guess," Eugene said.

"Only natural," said Pete. "You own him."

Again the tired lifting of the gray, fleshless mouth. "Yes, I do, don't I?"

They had turned the northern edge of Audubon Park past Loyola and Tulane universities and were headed now toward the zoo and the finish of the five miles they ran every morning. As he always did, when they came into the last quarter-mile, Eugene went up on his toes to sprint and cried, loud enough to turn the head of an old man walking twin poodles: "Balls! Who's got'm?"

But he knew who had them. Jacques had balls enough for four fighters. And despite the fact that he had already gone over four miles and outweighed Eugene by thirty pounds and had heavy boots on his feet, Jacques began to pull away from him in the sprint to the finish. And he was laughing all the way, his breathing as easy as if he had been walking. Eugene pulled up, leaned over, and put his hands on his knees, gasping. Jacques came to him and put his huge hand on the back of Eugene's neck.

"We run some, us, huh? You rat dere, leetle wan, all de way. Seet down you wanna, huh?"

"I don't need to sit down. I got your sit-down hanging, youngun."

"Youngan? Leestin to de man go wan. You neider hardly be older dan me, you."

"Old enough, son. Old enough."

Jacques laughed and bounced on his toes, hooking and jabbing, as they walked slowly past the zoo, cooling out, headed toward the apartment and a shower. As he always did, Eugene glanced toward the entrance to the zoo and thought of the lions, thought of them pacing the little stream there under the inconstant gaze of the tourists hanging over the railing at the top of the impossibly high stone wall of the cage that was not a cage but their natural habitat that they could never leave, not even for an instant. But, Eugene thought, if he had gotten out of his own cage, nothing was impossible. All the lions

225

needed was a little help and a little luck. Like a goddam nuclear bomb being dropped dead center on the city, leaving the stone cage cracked open like a rotten walnut. Eugene smiled. Yeah, help and luck. When a fucking nuclear bomb was help and luck, you in a real trick of shit.

"Mizz Charity, she's not . . ."

"No," said Eugene, "she's not. Be cool."

They were approaching the apartment on Exposition Boulevard, and Jacques was only making sure that Charity was not there before he went in to shower.

"I could jomp down my apartment an . . ."

"I don't want you to jump down your apartment, Jacques."

"Don make de fun weeth me, you."

"Don't ask me every morning, then. One morning she'll be here and it still won't matter. Charity likes you."

"I like Mizz Charity, but . . ."

"But what?"

He looked off toward the horizon and breathed deeply. "I don know."

"You're not the first person Charity's made nervous, man. Don't worry about it."

The second night Jacques was in New Orleans, Charity had given a small catered party for him, which Eugene had thought a silly idea.

"A party for a fighter? He needs a lot of things, but a party is not one of them. Or if you got to do it, wait until he's won another fight, then have one for him."

"A small party, Eugene. It's not a lot to do and it would please me to have it. Just a few friends. Pete and Tulip, of course, and Jake, maybe Femally . . . Do you know her?"

"No."

"Friend of Jake's. She's nice, you'll like her. And maybe Jacques has a girl he'd like to bring. It'll be fun and make him feel more at home."

"He doesn't need to feel at home. He's a fighter."

"So you keep saying. Don't be such a hardass about everything. Now that things are going your way a little, try to relax and enjoy it."

"Christ, he's two days in town and he's going to a party. It ain't the way to handle a fighter. I wish you'd do me—and Jacques—a favor and stay out of his life."

"What in the world's got into you? You don't seem like yourself at all."

"Who do I seem like, then? Never mind," he said, before she could answer. "Have the party if it makes you happy."

"It makes me happy. It'll make everybody happy. Except maybe you. Your trouble is you never learned to have fun. I know Jacques'll have fun."

But Jacques did not have fun, and Eugene knew that he would not from the moment he walked in. This was not the place you wanted your fighter. These were not the people you wanted your fighter with. Platters stacked with food were set about the apartment and open bottles of wine seemed to be everywhere and there were fresh-cut flowers and soft music played on the stereo. Eugene had gone to pick up Jacques from the place he was staying until Pete and Tulip moved out of their apartment on Prytania Street, and since he was in no hurry to get there anyway, the party was well on its way to being drunk and glassy-eyed by the time he arrived. Pete was there with a whiskey glass in his hand and his good-time look on his face, as though he had just discovered somebody had died and left him a gold mine, and Tulip was with him, her eyes bright and the muscles in her jaws leaping under her fine skin as she ground her teeth from time to time. Jesus, she was speeding her brains out, and Eugene felt the depression that was already on him settle deeper.

Pete took his hand and said, already lifting a bottle and a glass from a table, "My man, my man, let me fix you up something for you blood." He tapped Jacques on the shoulder. "How you feeling, champ?"

"I don't think I need a drink," said Eugene.

But Pete did not seem to notice what he had said and put a glass in his hand, and neither did he seem to notice that Jacques had not said anything but only looked confused, maybe a little upset. Then to Eugene's amazement, Charity came out of the kitchen leading Purvis by one of his pudgy hands, her other hand holding lightly to the elbow of a slender, dark young man who moved graceful as a dancer.

"Well, about time," she said. "Where have you kept the guest of honor?" Her face was bright with an alcohol flush and she kissed Jacques lightly on both cheeks, which seemed to startle him, and he took a small step backward. Jacques had not said a word since he came through the door and Eugene could feel the questioning look Jacques kept giving him.

"Eugene," said Charity, "this is Mr. Blasingame's man, Purvis Reeker."

Eugene said: "I know Purvis."

"But he doesn't know my man, Ernesto," Purvis said and gave a short, startled-sounding laugh. "Ernesto, this is Mr. Eugene Talmadge Biggs." He turned to Jacques. "And this pretty thing is Mr. Jacques Deverouge."

Ernesto showed brilliant teeth. "It is my pleasure." He offered his hand and Jacques took it reluctantly, as though he were being offered a bit of shit to hold.

Jake had sauntered into the room wearing a tailored dark suit with a pale blue shirt and a wide red tie. She was the only one Eugene had seen who looked like she still had her wits about her.

She smiled and said, "Good to see you, Eugene."

"Come, Jacques," said Charity. "I want to show you. A whole roast Cajun pig. In the kitchen. Done in your honor." She turned and headed for the kitchen, with Ernesto and Purvis following. Before he moved away, Jacques leaned toward Eugene and said through his teeth: "Who ees thees fat cocksuckair call me preety thing?"

Eugene shrugged hopelessly. "What you heard is all I know. Don't worry about it. He's harmless."

"But Jacques Deverouge ees not harmless." He walked away toward the kitchen where Charity was calling his name.

Jake was standing near the wall with one hand on her cocked, lean, and beautiful hip.

"How can you stand this sober?" Eugene said.

"I can stand anything," she said.

"You don't know what you're doing. Trust me, you don't."

"I haven't known what I was doing since I was eight years old," she said. Then before winking and walking away: "And I haven't worried about it, either."

In spite of himself, Eugene poured a glass of vodka and collapsed onto a loveseat in the farthest corner of the room. And that was where he was still sitting when he heard high laughter that sounded a little hysterical, and Charity's voice saying, "Oh, Jacques, how absolutely marvelous!"

Eugene set his glass down and went into the kitchen, passing Pete on the way, who only rolled his eyes to the ceiling and held his alcohol-frozen grin while Tulip ground her teeth, her eyes no longer shining but now stunned and disoriented.

In the kitchen Charity was standing in front of a table that held a dark pig with something in its mouth, perhaps an apple. Whatever it was was burned black. One of the hindquarters of the pig had been torn away, leaving what looked to Eugene like a wound in living flesh. Purvis, his mouth shining with grease, was gnawing the last bit of meat off a bone.

"How wonderfully ethnic," Purvis was saying, sucking at the bone.

Eugene did not know what "wonderfully ethnic" meant, and he didn't like the expression on Jacques' face. It looked dangerous.

"For God's sake, don't misunderstand," Charity said. "I do think it wonderful. I really do." She looked at Eugene. "His mother and father are named Tante and Nonc. God, this boy is a find. He truly is a find."

"I don unnerstan," Jacques said.

"I don't, either," said Eugene.

Purvis reached and tore a thick slab of meat from the hind-quarter of the pig and offered it to Jacques. "Here, press some of this to your face; you'll feel better."

"I don eat the pork," Jacques said.

"But you're Cajun," said Ernesto. "This is good Cajun pig."

"Fighters don't eat pork," Eugene said. He slapped Purvis's hand. "Git that piece of greasy shit away from him."

"Does everything have to be such a production?" cried Charity. She looked as if she might burst into tears. "I think Nonc and Tante are lovely names, I do. And I only got the roast pig because I thought it would please him."

"I'm sorry, Mizz Charity," Jacques said. "I dent . . ."

Eugene touched Charity on the shoulder gently, when what he wanted to do was take her by the throat. "Let me just talk to Jacques a moment." He led Jacques out of the kitchen.

"What we gone do now, us?"

Eugene never even looked back as he opened the front door. "We gitten the fuck out of here." Jacques followed him silently through the door and out into the night.

By the time Jacques and Eugene got to the apartment on Exposition Boulevard they had cooled out from the run. Inside, Eugene reached into a drawer and threw Jacques a towel. "You get a shower and I'll start some eggs and a steak."

"If you wan clim een the shower first, you, I kin . . ."

"Hey, I'm the trainer. You're the fighter. Get in there and get to it. You gone fight for me, you got to eat."

"Sho, now, Eugene. We gone be all right, us."

"Of course we gonna be all right, us."

"You makin the fun, agin, you."

"I hope sooner or later you'll learn to talk English. Coonass don't sound right."

"An cracker talk do?"

"You get the shower. I get the food. We eats good here. Mais yeah, cher."

"I be making a coonass a you erry day."

"Sure you will."

From the kitchen, with the eggs in the poacher and the steak under the broiler, Eugene listened to the water running in the bathroom and thought again of the night of the party. When he had come home long after midnight, the apartment was empty, the lights on, and half-eaten food and empty bottles everywhere. He undressed and turned off the lights and went to bed, concentrating to keep his thoughts on Jacques and the job ahead of him. There was no point in thinking about Charity. He could not save her—although what he would have saved her from he didn't quite know—but he could and would disentangle his life from hers if he was just deliberate enough, careful enough.

Then the sound of the shower had awakened him with first light coming through the windows of the bedroom. The shower finally stopped and he saw her, naked in the pearling light from the window, coming toward the bed, and he smelled the bath oil as she lay down beside him.

She was still a very long time before she said softly, "Are you asleep?"

"I was when you came in."

"I'm sorry I woke you. I tried to be quiet."

"Don't be sorry. Go to sleep."

"Are you still angry?"

"I ain't angry."

"You were terribly angry."

"I don't want anybody fucking with my fighter. Anybody. Much less a fat fag trying to feed him goddam pork."

"Purvis is a good human being. He can't help it if he's a fag."

"No, I guess not."

There was another long silence and then: "Eugene, please believe me, I'm sorry as I can be about the whole thing. I really am."

"It's all right."

"Nobody's going to mess with Jacques. Certainly not I. I made a mistake. I'm just terribly happy and being so happy made everything come out wrong, I guess. I'm so glad you took me with you on the boat. I've made a new friend, a great friend."

"Jake?"

"Yes, Jake. She's been marvelous to me. I think she'll be a real help to me in my work. She wants to help me with my research, and God knows I need her."

The mention of her research brought a little stab of fear. Her research. He refused to let himself think about what she might have in mind for Jake.

"Try to get some sleep," he said.

"We're still friends?"

"Yes," he said.

"And lovers?"

"Yes," he said.

Christ, after all this, after everything that had come down and he was still lying.

Jacques came in from the shower, dressed in street clothes now, his dark hair wet, and the color in his face still high from the run. Eugene had a place set at the table and was just taking the steak from under the broiler to put on the plate beside the poached eggs.

Jacques sat down and said, "I never eat a poached egg in my life till I meet you."

Eugene set a tall glass of orange juice beside the plate. "It's a whole hell of a lot of things you never done that you gonna do now that you met me. Get to it before it gets cold."

Eugene sat down across from him with a glass of juice but no food.

"You oughta eat in de morning, you. No good not to eat in de morning time."

"I never been much for breakfast, Jacques. I did it when I was fighting but I never liked it. You're the guy who's got to go in the gym today, not me. I don't need it."

As he often did in spite of himself when Jacques was eating and he was watching, he thought of Budd. He knew it was crazy but he even felt a little like Budd. He had a fighter now, a fighter with whom he almost lived and for whom he had profound respect and hope and affection. Yes, maybe even real affection. He was not much older than Jacques but looking at the boy across from him working on the steak with great satisfaction, he felt very old indeed. Maybe it was because of all that he had hoped for himself, because of his own failed dream, and maybe most of all because of everything that had happened to him—that he had allowed to happen to him in New Orleans— that he felt that way. He didn't know. And it did not matter, anyway. What mattered was that he had somehow come full circle. That boy sitting across from him was himself. He had come full circle and was still alive, did not now feel defeated beyond the hope of some sort of victory.

And Budd could not help him now. At odd times he wondered where Budd was, what he was doing. He had even thought of trying to get in touch with Budd, tell him he had a fighter, invite him into the deal Pete and he had with Blasingame. But it was no more than a thought, a kind of fantasy like the one he had of going back to work the land with his daddy and brother. God knows they could use Budd to help them guide and train the kid. But it wouldn't happen. Not Budd nor anyone else could help him now. He had to do it himself. He knew it and accepted it. But he needed somebody and he had nobody. He felt at times as though he were out of his depth, as though he had stepped through a door and found not footing on the other side but an airless void. He knew that Pete was responsible in part if not entirely for this feeling. But he was not going to blame Pete. He had his own troubles. Even thinking of Pete, though, gave him great pain. There was something bad going on in Pete, with Pete, something very bad. He did not know what it was but it felt like betrayal, a kind of blood betrayal. And Eugene knew that it was bound up with Tulip, what and who she was, and with Blasingame, what and who he was.

He was deeply worried about Pete and that made him worry about himself or at least worry about whether or not what they were trying to do could succeed. Pete was drinking more than Eugene had ever seen him drink, and he did not seem interested enough in their immediate problem: making Jacques the fighter they would have him be.

And, too, Eugene knew that Pete had lied to him. It didn't anger him, it only made him feel more lonely and isolated than he felt already. Lonely and isolated and saddened to the point of tears.

"Say what?" Pete had said but he had not looked up from the drink in his hand.

"I said it straight enough," Eugene had said. And he had. "Tulip was speeding her brains out last night, her eyes turning sixes and sevens. At least they were before they went dead."

It was the morning after the party for Jacques, and Eugene had come to the apartment on Prytania Street, where he found Pete sitting amidst packed cardboard boxes of clothes and dishes, sitting on a ladder-backed chair with a glass of whiskey in his hand, the bottle on the floor in front of him. Tulip had gone out. Pete didn't say where. They were moving the next day to the building where Jake had her apartment.

Pete looked up at him. The whites of his eyes had turned,

gone to the dull color of cooked egg yolk. "You ought to slack back, Jack. Be my woman you talking about."

"Jesus, Pete, this is me, Eugene. We can talk about anything, caint we? Talk to me, for Christ's sake. You got shit in your life? Then I got shit in my life. If it was something I couldn't pull by myself, I'd come to you for help. You know I would."

"Ain't nothing in my life I caint pull."

"It's hardly even the middle of the day, and you sitting here drinking straight whiskey."

"I know what time it is," Pete said, picking up the bottle and splashing more whiskey into his glass. "Jus a little a hair, man, hair a the dog that bit me last night."

"Tulip gone to git a taste of the crank that bit her?"

"You pissing me off, and you messing in the wrong place with the wrong man. I told you, goddammit, that I had her clean. I say she clean, she clean. All right?"

It was not all right. What Pete had said he said in a voice so despairing and unconvincing that instead of talking anymore about Tulip, Eugene wanted to hug him. This was his brother. They had been to war together and survived. And now it had come to this. It had goddam come to this. And Eugene didn't know what to do. He wanted to make it right and did not know how. He just had to hope Pete was strong enough to do what needed to be done.

Eugene had put his hand on Pete's shoulder. "I'm sorry, man."

"For what?"

"I wish to fuck I knew."

Pete looked into his glass and said, "Jus because we got a business deal, don't think you my daddy."

"I'm gone let that go. It ain't nothing but whiskey talking, anyway. The only thing I ever thought was that I was you friend. That we was friends."

Pete looked up and the old smile was back on his face, but his eyes looked whipped. "Now ain't this some shit? Listen us talk." He put the glass on the floor and stood up. But he was not steady on his feet, and Eugene knew that he had to have been drinking on and off through the night. "All right," Pete said. "I been a little out of line. Believe it, bro, I'm O.K. Tulip's fine. We gone be righteous. I ain't used to good luck and we fell into more than I ever had. We got our shit together though. Believe it."

Eugene had not believed it but he'd said, "I believe it."

Eugene watched Jacques slowly chewing the last bit of undercooked steak. He sighed, pushed away from the table, and stood up. "I'll grab a shower and we're out of here."

"Good," said Jacques. "Errythin good. Good day, good run, good food."

"Yeah," Eugene said. "Everything's just fucking great."

When he came out, showered and dressed, Jacques had rinsed his dish and the glasses and put them in the washer and sponged down the table and the stove. They walked together out to Blasingame's lease car sitting in the garage beside the 1000cc BMW motorcycle. Jacques stood a moment looking at the bike.

"That ees wan fine machine."

Eugene glanced at the motorcycle. "Yeah," he said. "Yeah, it is."

"You don take me for a ride yet."

"Not likely to, either. We wouldn't have to drop that but once at speed and you ain't fighting no more."

"You ride her."

"Not when I was fighting, I didn't."

"But after you leave the ring you ride her good."

Eugene regarded the bike for a moment and then opened the door to the car. "After I left the ring, it didn't matter one way or the other whether I dropped it or not. Not at all."

On rainy days, after a run they went to the River Walk down by the Quarter and strolled among the hundreds of shops and talked and watched the Mississippi River, which the shops overlooked, and relaxed, letting the hours slip by until it was time to go to the gym where they would meet Pete and Jacques would spend the afternoon on the speed bag or heavy bag or in the blurring circle of the skip rope before he went a few rounds with a sparring partner and ended with steam and a massage and Eugene took him back to the apartment on Prytania Street and put him to bed.

But if it was not raining—and today was fine, under a brilliant, solid sky, the air not particularly hot but heavy with humidity—they headed north on Elysian Fields, out past the University of New Orleans, and walked along the beach on Lake Pontchartrain just down from an abandoned amusement park. The lake was immense, so huge that it was easy for Eugene to imagine that it was the ocean he was walking beside, except that there were signs everywhere warning against swimming in the calm, lovely cobalt water. The lake was probably poisonous, perhaps lethal for all Eugene knew, deadly with the human filth and waste that poured into it from every direction. On the lake on every day of every week, the white, pristine wedges of sailcloth stood in sharp relief against the sky. It made for a beautiful, relaxing thing to look at, and Eugene wondered how long it had been since a fish had been able to survive those waters. Maybe there were still fish in Pontchartrain, but Eugene thought probably not, and if there were, they were probably cancerous with chemicals, their scaly backs spotted with open, spongy ulcers. And if it wasn't so now, it was only a matter of time.

"Thees ees some very beautiful," said Jacques as they sat on one of the benches placed on the grassy strip just above the beach, the pleasantly warm air washing over them. "Some very beautiful."

"It is beautiful," Eugene said, "if you don't think too much about it."

"I don unnerstand."

"I don't, either. Never mind. Don't mean nothing. I just got about half a case of the redass."

Jacques opened his mouth to speak, didn't, but looked back out toward the miles of open water and the sails against the far sky. Then finally he did speak, his eyes still fixed on the wide water of the lake. "Eet ees some bad habit not to think. A man always should be a thinking, yeah. Daddy Nonc say a man ain't been thinking gone end up sho with a mocasockasin in his mouth."

"A moc-a-sock-a-sin, Jacques?" He smiled in spite of himself, feeling suddenly better about the day and what it might yet bring.

"A mocasockasin, sho now. Daddy Nonc say this. In you mouth."

"Well, I ain't wanting a snake in my mouth. Not even a moc-a-sock-a-sin."

Jacques turned his eyes on Eugene now, his strong dark face unsmiling and serious. "You makin the fun with me agin, you."

"I'm not, but I could straight-by-God use some fun. I could do that."

"Caint be fun all de time Daddy Nonc be telling me since I wee leetle boy."

"How bout some of the time, fun some of the time?"

"Sho, now."

"What would your daddy be doing right now, Jacques?"

"Back home now?"

"Back home right now."

Jacques Deverouge came from way back in some unpronounceable bayou south of Lafayette, the very heart of Cajun country. He often talked of his mama and daddy. He didn't want to do what they were doing, had left the bayou in fact to keep from it, but he had certainly never thought of abandoning them, and they were never far from his thoughts or, given the opportunity, his conversation.

"Oh, since before a long time now, Daddy Nonc in de pirogue, jus him and dat push pole running de traps for dem nutrias and muskrats, for sho now, all day long him and de push pole in de pirogue." Jacques waved his hand to include the whole horizon. "Daddy Nonc wooden like eet out here no way. Too open. Like de flat black water and de close trees, errythin hemmed in like. He don like to see him too far at no one single time, jus a leetle way to de nex bend in de flat water, him. He bitch and carry on, Daddy Nonc do, but he don come out dat bayou, no way. I hear him say more dan won time, 'I don know why I likes it out here. Trappin ain't nothin but blood an mess. But I don have nobody to bother me an I got all de food I needs right at me feets.' Damn strange, Daddy Nonc. An dem nutrias . . ."

As Jacques talked, Eugene settled lower onto the wooden-slatted bench and let his head back to rest and closed his eyes against the brilliance of the sun and thought how good it felt out here with the heat on the skin of his face and the heavy, warm breeze blowing off Lake Pontchartrain, and even though he had never in his life been deep into a bayou stand-ing on the narrow floor of a shallow-draft pirogue as it moved silently over the black water between banks of dense moss-shrouded trees, it was easy enough to imagine himself drifting there as he floated on the soft, mellifluous voice droning on, the voice itself lost in a kind of dreamlike reverie now, talking on about the hard and bloody work of skinning out nutria rats, not rats at all but big as beavers, yellow-toothed, feeders by prefer-ence on cane but able to live on practically anything, with hides that were not as profitable as muskrats but profitable enough for a Cajun, who like the nutria had foods of choice but could live on practically anything if necessary and it was usu-ally necessary, so Cajun trappers generally and Daddy Nonc specifically simply chose to ignore the rash he kept over most of his body, a rash called nutria itch, and went on with the unprofitable, bloody work of trapping, gutting, and skinning nutria rats because it left him in the bayou which, although it was a dangerous, dirty, and unlovely place, was his place and the only place he wanted to be in the world and so he would go on with the pirogue even though it was slow and wrung a man's shoulders to pole it all day, go on with the pirogue until his son won the world as a fighter and bought his Daddy Nonc a jon boat with a good dependable Mercury outboard which Daddy Nonc said he did not want but which he did want and would use with great pride when his good strong son won the world with fists and bought the boat and the motor for his daddy with love and for Mama Tante the good strong son with the same love would . . .

Chapter Fourteen

"**H**ey, you."

Eugene opened his eyes. The skin of his face was hot and drawn from the sun. "What?"

"Time we goin to de gym."

Eugene pushed himself up straight on the bench. His neck was tight from where it had rested on the top slat of the bench. "I went to sleep?"

"Like a baby."

"What time is it?"

"I don know, but time we goin to de gym."

"You should have woke me up."

"Any time a man kin sleep he oughta."

"Come on, then, let's get to it."

For two weeks now, since the third day Jacques had been in New Orleans, he had been training in a gym in Harvey, a little town on the West Bank across the Mississippi River. It was only a half-hour drive from Prytania unless it was rush-hour traffic, so it was convenient, a gym that Mr. Blasingame either owned outright or had an interest in. Eugene didn't know which, but fighters from all over New Orleans worked out there, and Jacques had all the sparring partners he needed, and the right sparring partners were the hardest thing for a fighter to come by when he was in training. Sometimes he needed speed work, sometimes power work, sometimes a fighter heavier than he was to go with, sometimes one that was lighter. In the gym, just off Lapalco Boulevard in Harvey, Jacques had all the work he could handle, and it was always work with the kind of fighter that Pete thought Jacques needed most. Because it was Pete who took care of that part of the kid's life, the training in the gym. Pete had, after all, very nearly been a champion, and so Eugene was happy to let Pete direct Jacques in the gym and to arrange—to have the final decision on—whatever fights he was to have as he progressed as a boxer. Eugene was the kid's eyes, ears, and conscience outside the gym. He got Jacques up in the morning, ran with him, fed him, talked to him, and put him to bed at night.

Going over the Greater New Orleans Bridge to Harvey, Eugene said: "I heard what you said, what you were talking about anyway, back there on the beach."

Jacques smiled and put one of his thick, square hands on Eugene's shoulder. "Sho, now, you done been hearing it." He laughed and shook his head. "Like a baby you sleeping."

"Then how do I know about nutria itch trappers have and you gonna buy your daddy a jon boat with a Mercury outboard and the stuff you gonna do for your mama Tante when you win the world?" He couldn't say what stuff Jacques was going to do for his mama because that was when he faded out. But it seemed important to Eugene to pretend that he had in fact been listening and not asleep. He did care about the kid and what he had on his mind, and it was important that Jacques know and believe that. A fighter who was at peace with himself and believed in the people around him—believed they cared about him—a fighter like that developed on a straighter line than one who felt isolated and alone. And goddammit, he did care.

"You hear in you sleep, you?" It was a question of mild astonishment.

"I do a lot of things in my sleep, Jacques. Some of my best work is done when I'm flat on my back and sound asleep. But I wasn't asleep back there on the bench."

"You wan crazy white man," Jacques said.

"Seems like to me you might be right."

It was early afternoon when they pulled off Lapalco and parked, but the macadam lot adjacent to the gym was already filled with cars. The building was long and low and without air-conditioning despite the New Orleans heat and humidity. At the south end of the building the two huge exhaust fans, as big as airplane propellers, did not seem to make much difference one way or the other, and the odor of sweat and liniment and canvas and leather was heavy on the air. Fighters, some dressed in sweats and some in shorts and T-shirts, grunted and banged heavy bags or skipped rope or stood in seeming trances before the blurred staccato of speed bags. The two rings, one at either end of the gym, were occupied with fighters wearing headgears who bulled and popped one another with snapping punches, and the sound of gloves on flesh was unlike any other sound in the noisy gym, flat, heavy, and somehow moist.

"Don look lak Pete get here," said Jacques.

"He'll be here. He'll be here by the time you dress out."

When Jacques came out dressed in his black, soft-soled fighter's shoes and shorts and sweatshirt, Eugene set him to work on the speed bag, first saying, "Pete'll be here. He ain't here now, it's good reason." Eugene wished he believed that. But he did not. This was the first time Pete had not been waiting for them in the gym, and it bothered him enough that he went over to Drexel—the only name the short, balding man who ran the gym went by—and asked him if Pete had been in. Drexel was standing beside the front ring with a towel around his neck, watching two lightweights.

"Not today," Drexel said, using the end of the towel to mop his face. He didn't look at Eugene, but concentrated on the fighters. One was caught in a corner. "Get out of there, asshole!" Drexel screamed. "Get out of there!"

"I don't guess Pete called, left any kind of message?"

"Don't know nothing about no message," Drexel said and then screamed: "Jesus fucking Christ! Time! Time!"

Eugene called Pete's apartment and got no answer. Well, he would find out soon enough. There was nothing to do but go ahead with the workout, but leave off the sparring, even though Pete had had Jacques in the ring every day except the weekends when the kid didn't train at all, every day since he had come down from Baton Rouge. Pete always arranged for the sparring partners and was always ringside or up in the corner to coach, sometimes going into the ring to demonstrate slipping a punch or to show how and why a right-handed fighter always had to keep his lead foot inside the lead foot of a left-handed opponent.

"So what we do?" asked Jacques, coming off the light bag.

"We work," said Eugene. "Pete's not here, but we don't need him to work. Go to the heavy bag and, goddammit, act like you eat that fucking steak this morning. Take it off the chain!"

Jacques went to the bag in a single-minded fury. But he always attacked it that way. He never really needed any pumping up. The kid loved to work. If anything, Eugene had to hold him back at times or his enthusiastic savagery would take him too far. But Eugene always talked to him as though he needed talking to anyway, because—again—he wanted Jacques to know how intensely he cared, that he was never just there watching.

And so, methodically as though he were chopping wood, but savagely, too, as though he were butchering a blood enemy, Jacques worked the heart out of the afternoon, moving from one part of the gym to the other, from the heavy bag, to sit-up crunches on the mat, to body bridges on the back of his head and the point of his heels, to shadowboxing in front of a tall mirror, moving tirelessly, always well within himself, his breathing regular and unforced. It was a great joy for Eugene to watch him work, but finally Jacques had had enough, though Jacques himself seemed unaware of it, never once—as was his habit—glancing at the clock.

"That's it, Jacques," called Eugene. "Go get wet."

Eugene was leaning on the apron of the front ring when Pete came in. Pete came directly to him.

"Where's Jacques?"

"Where you been, man?"

"Workout go all right?"

"Went fine. The kid's in the shower. Want to tell me where you been?"

"Doing a deal."

Eugene had heard dopers say the same thing, and it made him think of Tulip and see the syringe filling up with blood all over again. "Doing a deal."

"I got Jacques a fight."

Eugene straightened off the ring apron. "A fight? Just like that you tell me he's got a fight? Don't you think I could of heard about it before . . . Who's this fight with?"

"Man, you hearing about it now. It's just something that come up. I had to get on it and stay on it to make it go. Hell, I thought you'd be as happy about it as I am. Been busting my ass all day trying to set it up."

"I didn't say I wasn't happy. It's just . . . So who?"

"Desmond Juker."

"Don't know him."

"I do. A cruiser weight out of Baton Rouge."

"Cruiser weight. Jacques ain't ready for a cruiser weight, not unless the guy's a dog. Unless he's a real dog, I vote no."

Pete looked away from Eugene toward the little door in the back leading to the showers. "He ain't a dog and Jacques's ready. He's in great condition and it'll be a good show. Trust me, he'll be fine."

"I know what kind of condition he's in. That ain't the point. The point is Jacques ain't a cruiser weight."

"So he gives up three or four pounds."

"Closer to ten."

Pete turned and put his hands on Eugene's shoulders. "Listen, this is a great opportunity. Juker's a fine young fighter. Six and 0. But Jacques . . ."

"Six and 0?"

". . . can take him. I ain't saying it'll be a waltz. If it'd be a waltz, it'd be damn little reason to do it. Nobody knows Jacques' name. He takes Juker out and a lot of people will know who the hell he is."

"How many rounds?"

"Six."

"I was afraid of that. I felt that coming. Jacques's never been more than four."

Pete took his hands away, took a deep breath, and looked at the ceiling before he looked back at Eugene. "Hey, this is Pete you talking to. I know what the kid's had. And I know what he ain't had. He ain't had exposure. He needs exposure. This'll get it for him. I thought you believed in me, man."

Eugene did believe in him. Did. He wasn't sure he did anymore. He looked back toward the showers. "Don't say anything about this to the kid. Let me talk to'm. I'll talk to'm when I take him back to his place."

Pete watched him a long moment and then said: "Sure. If that's what you want."

"I just wish I'd seen this Juker work, if it wasn't nothing but in the gym."

"I've seen'm work. I covered every angle. I wouldn't a made the fight any other way. I thought that was my job, the understanding we had."

"Yeah," said Eugene, "that was the understanding."

"I didn't just get out of bed and do it like that." He snapped his fingers. "I touched every base I could touch. That's where I been. Went down to Baton Rouge, talked to Joe Rocker and the Bad Man. They both think the fight's a natural."

"Rocker and Bad both think it's a good fight?"

"That's what I told you, ain't it? A natural. Besides, it's gonna make both of us a little something extra on the side."

Just in the way he said it, Eugene knew that he did not want a little something extra on the side. "How's that?"

"Mr. Blasingame's gonna promote it at the ballroom of The World Under Hotel in the Quarter. We still . . ."

"Blasingame?" Eugene liked it less and less.

Pete held up his hand to silence him. "We still need a main event and two more fights on the undercard with the kid and Juker. But that's easy. Piece a cake."

"I thought we just had a fighter. When did you start promoting, too?"

"Damn, Eugene, you making me tired here, and I was already tired when I walked in. Try to listen, man. Blasingame's promoting it. It's his risk. You and me ain't risking a fucking thing. And it's a great chance for Jacques. Blasingame owns the kid, and he's the one holding the bucks, remember? He wanted to do it this way. I told him about Juker and we talked about it, and when the nickel landed, this is the side that come up. Jesus, please don't fuck with me over this. I got enough on my head already without you fucking with me over this."

"You talked to Blasingame?"

Pete sighed as though he were trying to explain something to an idiot. "Yeah, I talked to Blasingame."

"You didn't tell me."

"No, I didn't tell you. I talked with him more than once. I

247

don't have to like it, I only have to do it. Comes with the turf, part of my job, my end of the deal. Christ, Eugene, you want me to check in with you about everybody I talk to, is that what you want?"

"No, that's not what I want." This whole conversation with Pete had first depressed him and now sickened him. The tone of it sickened him. However it had been between them was destroyed. Or maybe it wasn't. Maybe that was the wrong way to think. They had never had a fighter together. Things had changed up. Maybe that was it. Pete was no longer running snuff films for tourists, and he, Eugene, was no longer up to his neck in kink.

Eugene turned to see Jacques swinging across the gym toward them in his long, easy stride.

"Hey, you, Pete," Jacques said.

"What it is, champ?" said Pete, throwing a light left hand which Jacques slipped and faked a hook underneath. "You git a good workout today?"

"I always git a good wan."

"I thought you'd be better missing the sparring today," Pete said. "Sorry I wasn't here. Some things come up I had to take care of."

"No problem, Pete," said Jacques putting his arm across Eugene's shoulders. "My man was here."

"You could do worse than have Eugene in your corner."

"I'm hearing dat."

Pete said: "I got to hustle out of here. Eugene, try to get up with me this weekend. Some things we need to talk about. But call first. Make sure you *call*. I may be out, all the shit I got coming down. You guys gonna be all right?"

"Got everything covered," Eugene said.

"Errything," said Jacques.

"Another time, another place then. Later on."

They stood watching him until the heavy gym door swung shut on the blaze of brilliant sun he disappeared into.

"Pete wan fast guy," Jacques said. "Movin alla time. Hurry an run, run an hurry."

"Yeah, he's fast, a fast guy. You ready to pack it in?"

"I stay on ready," said Jacques.

Their normal routine in the evening was to have dinner together in Jacques' apartment, all the while talking fighters, Eugene telling and retelling the endless stories he'd had from Budd, all of which fascinated Jacques, who had never heard any of them, and then they would go for a walk or sometimes to a movie. Eugene made the kid's life his whole day, every day, got him up and put him to bed at night. Budd had done it for him and impressed upon him that it was the only way to do it, that a fighter, for any number of reasons, needed that kind of close handling and attention, particularly when he was young. But back on Prytania Street, Eugene begged off. "You be all right if we don't hang out tonight? I need to be at my place."

"Sho, now. You stay away from dere too much. A preety woman take much a lot of time. Daddy Nonc is telling me all about dat."

"Don't go out fucking around. Just because tomorrow's Saturday and you ain't running or in the gym don't mean it's all right to go jacking around and lose sleep. Got it?"

"Got it, boss man. You tell Mizz Charity hello, you."

"Sure."

But it was not for Charity that he was going back to the apartment. Charity was gone most of the time; he hadn't even talked to her in four days. Which was fine with him. But he did want to talk to Joe Rocker and he couldn't very well use Jacques' telephone.

It was Rocker who picked up the phone in Baton Rouge, and Eugene knew it would be. Joe Rocker practically lived in the gym.

"This is Eugene Biggs, Joe. I didn't get you away from something, did I?"

"It's always something, son, but no, I can talk. What you got on you mind? How's Jacques?"

That question seemed strange, if Pete was just down there today.

"He's doing fine. I wanted to ask you about a kid named Juker."

"Desmond? Damn, you ain't in the market for another fighter, are you?"

Stranger and stranger.

"He any good?"

"A banger with both hands. Undefeated as a pro. Young, trains hard. Got a lot of heart. Not much experience as an amateur. Cruiser weight. But he'll grow into a legitimate heavy. Why you ask?"

There it was. Pete had told him a stone lie. "I know a promoter that might be looking to put together an undercard, that's all. Say, did Pete get by to see you when he was over there?"

"Ain't talked to Pete since he was here with you. Tell whoever you talking with that this kid Juker'll put on a hell of a show. Strong as a horse and he fights from bell to bell."

"I'll do that. Good talking to you."

"Give Jacques my best."

"Right."

Eugene put the phone down and sat very still for a long time. What now? What to do? He'd have to talk to Pete. Certainly he'd have to do that. Just get everything in the open. Eugene wasn't going to let him fuck over this kid, not him, not anybody. Had something already been signed committing Jacques to the fight? Did Pete have the power to sign without his consent? Did Blasingame? He didn't know. What a trick of shit. All of this with the kid maybe could have turned into a good thing, but now for sure and for certain it had turned into a bad thing.

Eugene went to the icebox and opened the freezer. A half-quart of Stolichnaya sat frosting on the ice. Eugene poured a couple of ounces and swallowed it. No sooner had it fallen on his stomach than he remembered he had not eaten today. But he did not feel like eating, had no hunger.

"It don't matter what you feel like," he said aloud. "You won't be any good for the kid starved to death."

He got out three eggs to scramble and an English muffin to toast, but before he even cracked the first egg, the hundred-proof vodka felt so warm and comforting in his stomach—almost like the company of a friend, and God knows he needed one—that he poured another drink and knocked it back.

By the time he got the eggs scrambled and the muffin toasted, the food lay on the plate like an absolute insult, because while he cooked, he had sipped steadily from the bottle and had caught something more than a buzz, something that was right on the edge of drunk. He told himself to eat the food anyway.

But he never got the chance. The doorbell rang. He stopped between the stove and the table, the plate in his hand, trying to think who it might be. He put the plate down and went through the apartment to the front room where no light had been turned on. He opened the door and through the screen he saw that it was a woman, but she was only a dark shadow standing there in the gathering dusk. Then white teeth flashed, and when she called his name he knew who it was.

"Femally? It is you out there, right?" He leaned nearer the screen door.

"That's right, white boy."

"You oughten to call me that. Sounds shitty."

"Shitty enough," she said, "white boy."

"You gonna be ugly and nasty and fuck with my head, you might as well come inside to do it." He reached to open the screen.

"I don't need to come inside."

He stopped with his hand on the door. "What is it you need?"

"I need for you to take care of your house so I can take care of mine."

"Take care of my house?" Was he drunk, or was she talking twisted?

"Charity." She said the word like an oath.

"What about her?"

"She messin in my shit."

"Probably. What you want me to do?"

"Take care of it."

"I don't care what she does. Ever."

"I do. I care a lot. And I won't come up on her like you come up on Dong. I'll come upside her head with my razor."

So she knew about Dong. He didn't care. That was over, history. He didn't speak but only stood looking at her. His eyes had adjusted now and he could see the beautiful black planes of her face.

"You know where Charity is?" she said.

"You mean right now?"

"Right now."

"I try not to think about it," he said and remembered Jacques and the danger of a snake in his mouth. "But she said she was doing research."

She gave a short, unpleasant laugh. "Is that what she calling it, white boy?"

"This is beginning to piss me off, Femally. And my supper's getting cold."

"That bitch be gittin cold she ain't careful."

"Either come in and talk to me while I eat or let's shut this down."

"I ain't coming in, but you better be going out. Take you a little ride over to her place."

"Not a chance," he said.

"Eugene," she said, and in saying his name, her voice had changed completely. It was at once sad, pleading, and utterly lost. "Eugene, help me, man. I don't want to hurt nobody. You go to Angola for that. My life ain't much but it's all I got. Go on over there. Take a ride. What can it hurt?" She paused, and if it had been anybody but Femally standing on the other side of the screen, the cocky, arrogant, cynical Femally that he remembered from the terrace in the Quarter, he would have thought that what he heard was a sob catch in her throat. "Please," she said.

Before he could think of how to respond, or even what she was really trying to get him to do, she turned from the door and was gone into the darkness.

He went back into the apartment with every intention of forgetting he had ever opened the door. But he couldn't. And the reason he couldn't was Jake. He didn't know Femally and while he didn't really know Jake, he and Jake had spent part of a long drunk night talking with each other when both of them were badly hurt. And he liked Jake. He thought she was good people, had thought so since that first night on St. Charles. He stood in the kitchen looking at the eggs cold on the plate. He poured another drink, a stiff one. What could a ride hurt? He'd shoot over to Napoleon, only a few blocks away, and see Charity, try to find out what was going on. He knew, though, what was going on. Jake was going on. Maybe he could do something. Maybe, too, you jackleg, he thought, that's the vodka talking. But whatever went down, it would only take a minute. He owed Jake a minute.

The houses on Napoleon Street were all ablaze with the light in the dusk of early evening. Charity's apartment, though, was dark, and he almost drove past, but there at the very back he saw a faint light in a small window. He swung to the curb and went up quickly to ring the bell. He wanted to get this— whatever this was—over in a hurry. He had his hand raised to the bell when a sound stopped him, a lilting, drawn-out sound that reminded him of a sound of his childhood: nesting doves. He lowered his hand and waited to make sure that he had actually heard what he thought he had. And it came again, long and lilting now, a sound only a woman could make. But it was not Charity, and while he knew it was a woman, it was unlike any sound he had ever heard a woman make. He backed away from the door and stood on the edge of the porch a moment, listening. What now? Nothing now. He'd go back home and try it again—if he tried it again—tomorrow. But without moving from where he stood on the darkened porch, he knew without having thought about it that the lighted window was Charity's bedroom, and against his will he remembered the night that seemed now a lifetime ago, standing at the foot of her bed, she sliding out of her clothes, slipping into nakedness, and he, all in a panic, striking himself into unconsciousness.

And then without thought, as though his senses had a will of their own, he was drawn off the porch, down the steps, and along the narrow aisle of grass running beside the house, back, on back, to the low, faintly lighted window. The crooning wail came again, louder now and more urgent. A narrow seam of brighter light no more than an inch wide ran across the bottom of the window where the blind was not completely down. He stepped to the window and stopped there.

Charity and Jake were naked on the bed, cast long and beautiful in the guttering light of two candles. Their limbs were joined and locked together. Charity's mouth moved between Jake's young girl's breasts, first over one and then the other. Jake's slender neck arched and her mouth opened and the sound, full of the grief of need, came again. Transfixed, Eugene watched. It was the most beautiful thing in the world; it was the most dreadful thing in the world. Charity's mouth moved up from Jake's breasts and caught the sound with her tongue, stopped it there with a kiss.

Then one of Charity's hands that had been holding Jake's

hips moved out, away from her hips, moved slowly toward the edge of the bed, and out still farther. Eugene watched the hand moving there at the end of the long, lovely arm as though it were the moving head of a snake, moving still, reaching, searching, searching tentatively. Then Eugene saw for the first time what the hand was searching for, when it crossed the little table at the side of the bed and touched the switch on a Dicta-phone.

Chapter Fifteen

Eugene heard the key in the lock sometime before midnight and knew it was Charity, but he bent to one of the three suitcases open on the bed, a matched set of luggage he had acquired since leaving the Hotel Rue, and put a folded shirt in it. He did not even look up when she came into the room.

"What are you doing?" she asked.

"Packing. Just packing."

"Well, I can see that. But what for? What's going on?"

"A fighter's going on, Charity. I thought you'd heard. I've got to move in with Jacques. I can't work it long range. I should have known that." He didn't mind the lies at all. He didn't mind anything now. He only knew that he had to get away. And he wanted to do it with as few problems as possible. The last thing he wanted was quarrels, recriminations, attempted cover-ups, or all-out confessions. He thought: Just let me get out of here, Lord, just let me get away from here. Tonight.

"You might have said something."

"I only decided today. And besides, I ain't seen much of you."

"I've been busy."

"Both been busy, we have. Training a fighter takes almost as much time as research."

"My research has been going really well. That's why I'm here."

Did Jake have to go somewhere and turn a trick? Or maybe they had only exhausted each other.

She said: "I thought we could have a nice night together."

He straightened up from the suitcase and looked at her and then beyond her to the black Dictaphone that had not been used since he and Pete had brought Jacques to New Orleans. Of course, she had had Jake to take up her time, and maybe that had been enough. Or maybe she had Dictaphones beside beds scattered all over New Orleans. The thought had occurred to him more than once. How utterly goddam depressing. He sighed, looked away from her, and dropped a pair of socks into the suitcase.

"I guess," she said, "that means we can't have a nice night together."

"I got to get this done tonight. I really am jammed for time."

"O.K., not tonight. But soon. I need to . . . Well, we need to finish what we started. That was the deal we had. Right?"

He thought of all the pages of his most secret life neatly filed in the other room and said: "Oh, I'll be back. Count on it. We'll have a great night. I promise you a great night."

"I knew I could count on you."

"You can, Charity. You can always count on me."

"Good. If that's clear, I don't mind you moving over to

Jacques'. Actually, I'd like to get to know him better. Because of you, I've found out just how fascinating fighters are."

Did she have a bed and a Dictaphone waiting somewhere for Jacques? He snapped the suitcases shut and set them beside the bed. She came to him and put her arms around him, her head on his chest. "You're sure you're coming back?"

"Now you're being silly," he said, taking her gently by the shoulders and pushing her back to look into her eyes. "There is no way I would not be back. Probably a lot sooner than you think." He reached in his pocket and brought out his apartment key. "It's all right if I keep this?"

"I'd be offended if you didn't keep it." She put her hand on his hand and closed it over the key. "Keep it. Use it."

"Then I'm out of here. Sorry to be in such a rush, but—as we say in South Georgia—when the ox is in the ditch, you've got to pull him out."

"You and South Georgia," she said.

"Yeah, me and South Georgia."

"Since I'm already here and I'm so bushed, I'm going to get in our bed and go to sleep. But it won't be the same without you."

"Jacques's a pretty thing," he said, "but he can never take the place of you."

"Don't," she said.

"Don't?"

"Joke about that."

"That's all it was, a joke."

"It's not something you ought to joke about."

"If you say so," he said.

She was already turning the covers down on the bed when he left, after first accepting her long kiss full of false passion.

Jacques' apartment was dark, but Eugene did not expect otherwise. He had a key, the same one he used to wake the kid up every morning. He already had the suitcases up the outside stairs and inside the apartment before Jacques came naked, rubbing his eyes and yawning, into the little room at the front of the apartment, which held only a couch and a ladder-back chair, all that Jacques said he needed.

"What is it?" he said, looking at the suitcases.

"It ain't nothing. Go back to sleep."

"What ain't nothin?" said Jacques. "Middle of de night. You suitcases here on de floor. Dat ain nothin?"

"Nothing we caint handle, right?"

Jacques looked at him, then the suitcases, and back again. "You got de women blue troubles, you?"

"I can sleep here, caint I?"

"Sho now, Eugene."

"Then snatch a mattress off your bed and throw it in here. We'll work something else out tomorrow. But it's late, I'm tired. I want to go to sleep."

Eugene wanted many things, but sleep was not one of them, and he knew it. He just wanted Jacques back in bed and the light out, so that the thoughts that pounded and tumbled about in his head could sort themselves out.

But they did not sort themselves out. Pete had lied to him, a stone, through-the-teeth lie. And Jake was being revealed to herself. If she could openly talk about driving young boys to suicide, what secret horrors would she give up to Charity and the Dictaphone in a frenzy of passion? And the umbrella under which all of them seemed to move was Mr. Blasingame. Oyster Boy. He owned the building where Jake lived, and where Pete and Tulip now lived, too. He owned Jacques and was going to promote his first fight. What did a man who owned a building in the CBD and a yacht big enough to float a city want with an unproven fighter? Did he want the same thing Charity wanted, to make him give up his secrets?

The thick dark was smothering him as he tossed naked on the mattress there in the stifling little room. He got up and turned on a light. He had to get out, get down on the street, do something. He was not careful to be quiet. There was no need of it. Jacques was a dead man once he had closed his eyes again. So Eugene got into his clothes, found his shoes, hit the light switch, and let himself out.

He opened both car windows so the air would roar around his head and turned the radio up. The raging sound seemed to quiet the thoughts, the unanswerable questions, and give him a touch of peace. The traffic had thinned considerably, and he drove without direction, distracted, listening to the rush of the wind and the roar of the music.

He did not realize he was headed for Pete's place on Canal Street until he got there. But when he looked up and saw the Jean Baptiste Lemoyne Arms, he knew that despite everything else that had happened, that was making a war zone of his mind, it was Pete who had driven him out of Jacques' apartment and into the street. He wheeled the car to the curb and got out.

The big black man in the lobby touched his cap when Eugene came in. "How you doing, Mr. Biggs?"

"I ain't no good, Marcel. You any good tonight?"

"Bout halfway," Marcel said. "How's that Cajun boy comin?"

"Like you say, bout halfway."

Marcel had had twenty-five fights as a pro and won fifteen of them, and now at the age of fifty had ended here on the door of a fancy condominium, but he had never got over fights and fighters. He had helped Eugene and Jacques and Pete the day Tulip and Pete moved in.

Marcel said: "Soak him in salt water and feed him jolypeño peppers. That git his young ass goin."

Eugene waved and stepped into the elevator. When he rang the bell on the fourth floor it was Tulip who opened the door, and her eyes were dead as stones. She said nothing. Her stare went right through him and through the wall behind him and out over the city. He glanced down and saw the swollen, purple, little pocket of flesh on the inside of her arm at the elbow.

Eugene raised his eyes and over her right shoulder Pete,

coming out of the bedroom, stopped, barefoot and naked except for a pair of Everlast boxing trunks, and stared at him. Eugene put his hand out and caught Tulip by the shoulder and carefully moved her aside. Something inside him—his heart or maybe his very blood—went cold and heavy.

"I told you to call first," Pete said in a hopeless voice. "You didn't call first."

"I needed to . . ." said Eugene before his throat closed on his voice and shut it off as Oyster Boy, naked except for the same Everlast boxing trunks that Pete wore, came out of the bedroom on his hands and knees, a leather collar covered with steel studs fastened about his neck. One end of a leash was attached to the collar and the other end of the leash was in Pete's right hand.

"We were expecting Purvis with the oysters," said Oyster Boy.

"Aw, Pete," groaned Eugene.

"Close the door," said Oyster Boy.

But Eugene did not close the door. He remained rooted to the spot where he stood.

"Junk's expensive," Pete said, his voice, again, flat and hopeless.

"No," said Eugene.

"I had to have the money."

"No," Eugene said.

"Until I can get her clean."

"Go out or come in," said Oyster Boy, "but close the door."

"No," said Eugene.

"I can get her clean, man," Pete said.

"No," Eugene said.

"I have to. I love her."

"No," said Eugene, but he did not mean no, that dope was not expensive or that Pete did not have to have the money or that he could not get Tulip clean or that he did not love her. He only meant No. It was one of the few utterly clear moments in his life when he was able to say exactly what he meant, exactly. "No," he said.

He turned and leaving the door open raced not for the elevator but for the stairs which he took two at a time, down four flights and past a startled Marcel who opened his mouth to speak but Eugene was through the lobby and the glass door opening onto the sidewalk before Marcel could manage a word.

Off Canal, driving through the empty side streets back to Prytania, Eugene knew exactly what he was going to do, what he had to do. He had known it would come to this for a long time now, but the right time seemed never to be at hand. One thing or another had kept him from it, but the time was now, tonight. Nothing could keep him from it, and a great calm settled over him, had settled over him from the moment he had said No. What a huge, warm cocoon of a word, wrapping and holding, serving to affirm and deny in the same instant. He felt as though he had crawled into No and pulled it in after him.

It took only a moment to go up to Jacques' apartment and find the .38 revolver. He didn't even need a light. He unsnapped the suitcase, slipped the .38 out of its holster, and put it behind his belt. He went downstairs and got in the car and headed toward Exposition Boulevard, driving slowly, as calm in his heart as sleep. On the way, he stopped at a twenty-four-hour Eckerd Pharmacy and bought a can of lighter fluid, a large can.

He opened the door of the apartment with his key and walked into the bedroom and stood a moment listening to Charity's soft, regular breathing. He slipped out of his clothes and stood naked at the foot of the bed holding the .38 in his left hand. The breathing missed a beat, followed by a rustling of the sheets, and when the breathing began again it was not so shallow.

"Charity," he said, softly.

"Eugene?"

"Yes."

"You've come back."

"Yes, I've come back."

He slipped under the light blanket and lay on his back. He carefully placed the .38 on the bed beside him. She moved to touch him. "You couldn't stay away."

"No, I couldn't stay away."

"Not even for one night."

"Not even for one night."

"Oh, you darling. You precious boy."

"Yes, precious." The steel of the pistol was cold against his leg.

"You had to come back," she said.

"Yes, I had to come back."

"To your Charity."

"To you."

Her hand moved on his stomach. The hard, heavy pistol lay on his palm, and he could feel the pulse beating in his thumb where it pressed against the steel.

"Do you want me?"

"I intend to have all of you. Ever last ounce of you."

He felt her breath catch in her throat. "Yes. Oh, yes."

"Open your mouth," he said.

"Knockout." She breathed the word. "My precious Knockout."

"Open your mouth."

She put her open mouth against his chest. Her wet lips moved across his stomach. She touched him with her tongue. He put his left hand on the back of her head, felt her finely shaped skull through the thick hair, let his hand slide forward over her flat cheekbones and thin nose, found her open mouth and long, trembling tongue.

His right hand came over his hips and he roughly pushed the barrel of the gun between her lips and into her throat and with his left hand behind her head, held her there. Her whole body shook violently and then stiffened.

"You dead," he said. "You just ain't been buried yet." She moved her head against his hand. "Be still," he said and was amazed at his own calm, flat, utterly violent voice, a voice he recognized as the voice of irrevocable decision. He was prepared to do anything, knew it, and accepted it. He took his left hand from the back of her head. "Give me your hand," he said. Her hand came to his. He held it a moment, then moved his palm gently across the back of it. He brought her hand to the pistol, made her touch the exposed, cocked hammer and finally put her finger on top of his finger over the trigger. "Steady," he said, "don't shake yourself into you goddam grave. If you do it, don't do it by accident. I'm here to help you. Press my finger. Just a little. Go on. You won't feel a thing." She was crying. Her whole body was trembling, even the finger that touched his finger that touched the trigger. He heard her teeth grind against the barrel of the gun. "You not gone do it? That's all right, I'll help you out. But first it's something we got to do. Reach behind you there and get the light." She did and the first thing he saw when the light came on was her wild and starting eyes looking up at him. "I'm gone take this thing out of your mouth. You can have it any way you want it. But the minute you don't do what I tell you, or the minute you scream, that's the minute you die. Death. We gone solve that great big problem for you."

He withdrew the barrel. The metal was wet and caught the light. A spot of blood grew on her swollen lower lip. She touched the spot with her tongue, but did not move.

"Please, Eugene."

"Shut up. Don't talk. Besides, Eugene ain't here. Too late for Eugene. This is Knockout. Get off the bed and stand up."

She got off the bed, her legs unsteady. One of her hands moved to cover the place between her legs, a gesture he had never seen her make before. Tears were running on her cheeks, but her stunned face was without expression. "May I get my clothes? A robe, any . . . ?"

"I think you can do this naked," he said. "And I told you. Don't talk."

A flush of blood started in her neck and washed upward, fanning across her cheeks. "You're crazy, Eugene," she said through clenched teeth.

"Knockout," he said. "Knockout was—is—crazy, yes. But what's crazy? Right? So shut up and listen." He had been sitting on the bed in front of her. He rose and stood beside her. He placed his left hand lightly on the nape of her neck and touched the crusting spot of blood on her lower lip with the barrel of the pistol. "You want to do this with the gun in your mouth? Is that what you want?"

Her eyes went wild again when he touched her mouth with the gun. She shook her head.

"Good," he said. "You're gone go in there and open that padlock, get all the folders, the tapes from the Dictaphone, the disks from the word processor"—all the time he was talking she was shaking her head, more and more swiftly until her hair was whipping about her face and she was racked with crying, a wet, choked sobbing—"and then you gone take it all to the kitchen and put it in the sink." She was wildly shaking now, her hands clenching and unclenching, totally out of control. He put his left hand in her hair and steadied her head and gently parted her lips with the pistol. "Put your tongue in it." She put her tongue in the barrel and some of the shaking left her. "Now suck on it." She sucked on it and quit shaking entirely. "Ain't nothing like sucking on the barrel of a gun to calm your nerves." He took the .38 away from her mouth. "Now let's get to it."

Her shoulders slumped, her head bowed, she did as he directed, the tears dripping off her chin onto the papers. It took three trips from the cabinet to the kitchen before everything was stacked in the sink.

He uncapped the lighter fluid and handed it to her. She looked at it for a long moment and then raised her eyes, no longer wild but defeated, broken.

"Use it," he said.

"For God's sake, don't. I'll give anything . . ."

"Ain't nothing you can give. Use it."

With her lower lip caught between her teeth she poured fluid over everything in the sink.

"All of it. Use all of it." She emptied the container. He handed her a box of matches. "Light it."

"I can't," she said. There was blood from her bitten lip on her teeth.

"I imagine you can." He lifted the pistol to her mouth and a thin stream of brown liquid slid down her legs. "I'm glad to see you'd sooner shit than die. Gives everything a little balance." He pulled her two steps back from the sink. "Light a match and throw it."

She did, and a ball of fire leapt halfway to the ceiling. He watched it as she sank to the floor. The papers curled. The dark smoke of the tapes and disks ballooned upward. Eugene watched it still. But it did not satisfy. None of it satisfied.

From the floor, Charity cried, "This is horrible, horrible."

"Yes, it is, Charity," Eugene said, watching the fire. "It is horrible."

He looked at her on the floor and felt no pity, felt nothing. He had been brought—allowed himself to be brought of his own free will—to the very face of horror. And now that he had broken through to the other side, he found nothing, a numb, frozen, featureless landscape of nothingness. He felt forever burned clean of feeling. He could have easily killed her there where she lay on the floor. But killing her and not killing her were balanced so evenly on the scale that he quickly slipped back into his clothes where he had dropped them at the foot of the bed, put the .38 behind his belt and walked out the back door into the early morning light.

Eugene rolled the BMW out, cranked it up, and eased into the Saturday morning traffic. He rode slowly, patiently, every decision already made, until he saw a used-car lot with a pickup truck on it, an old truck that looked like it had been taken care of, the sort of truck he had been thinking about. He wheeled into the lot and shut the motorcycle down. He walked over to the truck. Good rubber. Body solid without any evidence of rust. He raised the hood. One glance at the motor told him this was probably the truck.

"Know a cream puff when you see one, don't you, boy?"

Eugene turned to see a thin man with a substantial belly, wearing a yellow shirt decorated with palm trees, coming toward him.

"You got an ignition key?" asked Eugene.

The man held up a key. "Brought it with me from the office when I seen you walk straight to this little baby." He patted the truck's fender affectionately. "I said to myself, I said, Earl, that boy right there knows your basic cream puff when he sees it."

"Save that for somebody else, and give me the fucking key."

The man's open mouth clamped shut, but he never lost his smile. He handed Eugene the key. Eugene fired it up, let it idle, then held the accelerator halfway to the floorboard for a moment before turning it off.

"You want to drive it, you can . . ."

"I don't need to drive it. How much you give me in trade for that motorcycle?"

The man glanced at the BMW. "We ain't in the tradin bidness here, fella. Cash up front is the deal we make."

"I got registration, a clear title, and all the identification in the world. I need the truck today. Now. You want to talk, or you want me to roll on down the street and find somebody who does?"

"I don't think we can do bidness. But come on in the office. It won't hurt nothin to talk. Hot'n hell out here anyhow."

They went into the office, and an hour later Eugene drove off the lot in the pickup truck with three hundred and ten dollars of Earl's money in his pocket. When he walked into the apartment, Jacques was sitting at a little table eating his breakfast. He was just raising a loaded fork to his mouth when he looked up and saw Eugene. He put the fork back on his plate.

"What's wrong?" he said.

"Nothing's wrong. Everything's right. But I'm leaving."

Jacques went back to eating. "Too much you in an outta here too fast, you. What you got doin today?"

"I told you. I'm leaving."

"I heard. I mean what . . ."

"Town. I'm leaving New Orleans."

Jacques pushed his plate away. He opened his mouth to speak, closed it, then looked around the room as though there might be others there listening he could appeal to. Finally: "I don unnerstand."

"Nothing to understand. I swapped my motorcycle for a pickup. I'm putting my stuff in the truck and the truck in the road." He held up his hand to stop Jacques, who had stood up and was about to speak. "I'd explain but I caint explain. Too much, too twisted, Jacques. Listen to me now. Listen close. You in a trick of shit here. You got to get out, too. Everbody you've met, everbody, Pete, Charity, even Tulip, is bad, bad news from the ground up. Some of it ain't their fault. That ain't the point. They locked into whatever they locked into, but you ain't locked into it. Yet. But you will be. Believe it. Get out. Away. Now. I don't care where you go, just go."

Jacques smiled, visibly relaxed. "Then it ain't no beeg thing. I'm gone go wit you."

"No," said Eugene.

"Sho, now. I trus you, Eugene."

"Don't trust me."

"You tell me you do me bad? Tell me dat. Tell me dat an I forget all about you and where you go, you."

268

He held Eugene's eyes with his own, but Jacques was blinking rapidly now. This big, strong kid, who never blinked at blood, looked as though he might cry, and Eugene realized without surprise that he did not care. Eugene's heart was cold against the world. Nothing could touch him. Whatever he had killed back there in Charity's kitchen, and before that in the doorway looking at Pete, was dead. Still, he knew it was not in him to do Jacques wrong.

"I can't tell you that," Eugene said.

"Hell, I known dat," said Jacques, and he remained without moving, watching Eugene.

All right. If that's the way it is, that's the way it is. But he had to try one last time. "I got nothing, Jacques. No place to go, really. No plans. No money. No future. You come with me, we got nothing."

Jacques said: "We got somethin."

"What we got, Jacques?"

Jacques held up his fists. "We got dese."

The two of them stood for a moment, balanced in each other's direct gaze. Then Eugene said: "How long it take you to get ready?"

"Ten minutes too long, boss man?"

"Just right," Eugene said.

269